Measure Theory
and Probability

The Wadsworth & Brooks/Cole Mathematics Series

Measure Theory and Probability

Malcolm Adams, Ph.D.
University of Georgia

Victor Guillemin, Ph.D.
Massachusetts Institute of Technology

Wadsworth & Brooks/Cole Advanced Books and Software
Monterey, California

Wadsworth & Brooks/Cole Advanced Books and Software
A Division of Wadsworth, Inc.

Printed in the United States of America

10   9   8   7   6   5   4   3   2   1

Library of Congress Cataloging-in-Publication Data

Adams, Malcolm Ritchie.
    Measure theory and probability.

    Bibliography: p.
    Includes index.
    I. Measure theory.   2. Probabilities.
I. Guillemin, V., [date]   .   II. Title.
QA273.A5385   1986   515.4'2   85-26443
ISBN 0-534-06330-6

Sponsoring Editor: John Kimmel
Editorial Assistant: Maria Alsadi
Production Editor: Michael G. Oates
Manuscript Editor: Trisha Cain
Permissions Editor: Mary Kay Hancharick
Art Coordinator: Judith Macdonald
Interior Illustration: John Foster, Cool, California
Typesetting: ASCO Trade Typesetting, Hong Kong
Printing and Binding: The Maple-Vail Book Manufacturing Group, York, Pennsylvania

To Jon Bucsela in memoriam

Preface

Probability theory became a respectable mathematical discipline only in the early 1930s. Prior to that time it was viewed with scepticism by some mathematicians because it dealt with concepts such as *random variables* and *independence,* which were not precisely and rigorously defined. This situation was remedied in the early 1930s largely thanks to the efforts of Andrei Kolmogorov and Norbert Wiener, who introduced into probability theory large infusions of measure theory. In retrospect, it was fortunate that the kind of measure theory they needed was already available; it had, in fact, been created some thirty years earlier by Henri Lebesgue, who had not been led to the invention of Lebesgue measure by problems in probability but by problems in harmonic analysis. It seems strange that it took more than 30 years for this fusion of probability and measure theory to occur. In fact, since that time, probability theory and measure theory have become so intertwined that they seem to many mathematicians of our generation to be two aspects of the same subject. It also seems strange that the basic concepts of the Lebesgue theory, to which one is naturally led by practical questions in probability, could have been arrived at without probability theory as their main source of inspiration.

Saddled as we are with the fact that the theory of measure didn't develop along these lines, this doesn't mean we cannot teach the subject as if it *had* developed this way. Indeed, we believe (and this is the reason we wrote this book) that the only way to teach measure theory to undergraduates is from the perspective of probability theory. To teach measure theory and integration theory without at the same time dwelling on its applications is indefensible. It is unfair to ask undergraduates to learn a fairly technical subject for the sake of payoffs they may see in the distant future. On the other hand, the applications of measure theory to areas other than probability (e.g., harmonic analysis and dynamical systems) are fairly esoteric and not within the scope of undergraduate courses. Of course probability theory, taught in tandem with measure theory, is also not thought of as being within the scope of an undergraduate course, but we feel this is a mistake. *Discrete* probability theory is taught at many institutions as a freshman course (and at some high schools as a senior elective). The kinds of problems we will be interested in here, i.e., the amorphous set of problems that go under the rubric of the *law of large numbers,* are lurking in the background in these discrete probability courses, and are often so bothersome to bright students that they arrive, unaided, at quite original ideas about them. By formulating these problems in measure theoretic language, one is often doing little more than vindicating for undergraduates their own intuitive ideas and, at the same time of course, convincing them that the measure theoretic methods are worth learning.

By now we have probably given you the impression that this book is basically about probability. On the contrary, it is basically about measure theory. Sections 1.1 and 1.2 nominally discuss probability, but primarily discuss why measure theory is needed for the formulation of problems in probability. (What we hope to convey here is that had the Lebesgue theory of measure not existed, one would be forced to invent it to contend with the paradoxes of large numbers.) Section 1.3 deals with the construction of Lebesgue measure on $\mathbf{R}^n$ (following the *metric space* approach popularized by Rubin [see References, p. 199]. In §1.4 we briefly revert to probability theory to draw some inferences from the Borel-Cantelli lemmas, but §§2.1–2.5 are straight measure theory: the basic facts about the Lebesgue integral. Only to illustrate these facts do we return to probability at the end of the chapter and discuss expectation values, the law of large numbers, and potential theory.

Sections 3.1–3.5 are also consecrated entirely to measure theory and integration: $\mathcal{L}^1$, $\mathcal{L}^2$, abstract Fourier analysis, Fourier series, and the Fourier integral. Fortunately, the last two items have some beautiful probabilistic applications: Polya's theorem on random walks, Kac's proof of the classical Szëgo theorem, and the central limit theorem. With these we end the book. All told, taking into account the fact that we have packed quite a few applications to probability into the exercises, the ratio of measure theory to probability in the book is about 5 to 3.

The notes on which this book is based have served for several years as material for a course on the Lebesgue integral at M.I.T. and for a similar course at Berkeley. They have been the basis for a leisurely semester course and an intensive quarter course and have proved satisfactory in both (though in using these notes in a quarter course, we had to delete most of the material in §§2.6–2.8 and §§3.6–3.8). We divided the book into the three chapters not just for aesthetic reasons, but because we found that in teaching from these notes we were devoting approximately the same amount of time to each of these three chapters, i.e., four weeks in a typical twelve-week semester course. Incidentally, we found it very effective, for motivational purposes, to devote the first three class periods of the course to the material in §§1.1 and 1.2, even though in principle this material could be covered in a much more cursory fashion. We discovered that with these ideas in mind, the students were much better able to endure the long arid trek through the basics of measure theory in §1.3.

We would like to thank Marge Zabierek for typing the notes on which this book is based, and we would like to thank our students at Berkeley and M.I.T., in particular, Tomasz Mrowka, Mike Dawson, Harold Naparst, Mike Conwill, Christopher Silva, and Ken Ballou for suggestions about how to improve these notes and for weeding out what seemed to have been an almost endless number of errors from the problem sets.

We have dedicated this book to Jon Bucsela, to whom we owe an exhaustive revision of the manuscript before we had the final version typed. His untimely death in the spring of 1984 was a source of acute grief to all who knew him.

Malcolm Adams
Victor Guillemin

Suggestions for Collateral Reading

For background in probability theory, we recommend Feller, *An Introduction to Probability Theory and Its Applications.** We feel that at the undergraduate level, this is the best book ever written on probability theory. Its charm resides in the fact that there are literally hundreds of illustrative examples. This makes it hard to read through from cover to cover, but it is a gold mine of ideas.

Another beautiful book, though more advanced than Feller, is Kac's 96-page monograph in the Carus series, *Statistical Independence in Probability, Analysis and Number Theory.* Our treatment of Bernoulli sequences and the law of large numbers in §1.1 was largely borrowed from this book, and one can go there to find further ramifications of these topics.

There are several treatments of measure theory in conjunction with probability written for graduate students. In our opinion, the best of these is Billingsley's book *Probability and Measure,* which a bright undergraduate will, with a little effort, find accessible if he or she ignores the more technical sections toward the end.

Finally, for material on metric spaces and compactness, we have attempted to remedy the fact that we presuppose a nodding acquaintance with these topics by summarizing the main facts in the appendix. To learn this material, however, we recommend either Hoffman *(Analysis in Euclidian Space)* or Rudin *(Principles of Mathematical Analysis).*

---

*For complete bibliographic information for the titles listed here, see the Reference section on page 199.

Contents

Measure Theory
and Probability

Chapter 1

Measure Theory

§1.1 Introduction

In this section we will talk about some of the mathematical machinery that comes into play when one attempts to formulate precisely what probabilists call the law of large numbers.

Consider a sequence of coin tosses. To represent such a sequence, let H symbolize the occurrence of a head and T the occurrence of a tail. Then a coin-tossing sequence is represented by a string of H's and T's, such as

$$H\,H\,T\,H\,T\,T\,T\,H\,H\,T\ldots$$

Now, let s_N be the number of heads seen in the first N tosses. The *law of large numbers* asserts that for a "typical" sequence we should see, in the long run, about as many heads as tails. That is, we would like to say that

$$(1) \qquad\qquad \lim_{N\to\infty} \frac{s_N}{N} = \frac{1}{2}$$

for the "typical" sequence of coin tosses.

We do not expect this assertion to be true for all sequences, because it is possible, for instance, for our sequence of coin tosses to be all heads. Experience tells us, however, that such a sequence is not typical.

In what follows, we describe a mathematical model of coin tossing in which we precisely define what is meant by a "typical" sequence of coin tosses. With this model, the law of large numbers can be rigorously demonstrated.

Because James Bernoulli first stated the law of large numbers, in the seventeenth century, we will call a sequence of coin tosses a *Bernoulli sequence*.

Let $\mathscr{B}$ represent the collection of all possible Bernoulli sequences. Notice that $\mathscr{B}$ is an uncountable set. (See exercise 1.) This fact is also clear from the following proposition.

Proposition 1.   If we delete a countable subset from $\mathscr{B}$, we can index what is left by points on the real interval $I = (0, 1]$.

Proof.   We construct a map $I \to \mathscr{B}$ that is one to one and fails to be onto by a countable set. The map is constructed as follows.

Every $\omega \in I$ can be written in the form

$$(2) \qquad\qquad \omega = \sum_{i=1}^{\infty} \frac{a_i}{2^i} \qquad a_i = 0, 1$$

Because the a_i's determine ω, we introduce the notation

$$\omega = . a_1 a_2 a_3 \ldots$$

which is called the binary expansion of ω. From this representation we produce a Bernoulli sequence by putting an H in the nth term of the sequence if $a_n = 1$ or a T if $a_n = 0$. Unfortunately, this does not give a well-defined map $I \to \mathscr{B}$ because ω does not necessarily have a unique binary expansion. For example, $\omega = \frac{1}{2}$ has the two binary expansions

$$.1 0 0 0 \ldots \qquad \text{and} \qquad .0 1 1 1 \ldots$$

To avoid this problem we prescribe that, if ω has a terminating and a nonterminating expansion, we give it the nonterminating one.

This convention gives a one-to-one map $I \to \mathscr{B}$ that is not onto because it misses out on those Bernoulli sequences that end in all tails. Let $\mathscr{B}_{\text{deg}}$ denote the collection of Bernoulli sequences that, after a certain point, degenerate to all tails. We claim that $\mathscr{B}_{\text{deg}}$ is countable.

Proof. Let $\mathscr{B}_{\text{deg}}^k$ be the Bernoulli sequences that have only tails after the kth toss. Then $\mathscr{B}_{\text{deg}}^k$ is finite and

$$\mathscr{B}_{\text{deg}} = \bigcup_{k=1}^{\infty} \mathscr{B}_{\text{deg}}^k$$

is a countable union of finite sets. Thus $\mathscr{B}_{\text{deg}}$ is countable.   □

Because $\mathscr{B}_{\text{deg}}$ is a countable subset of the uncountable set $\mathscr{B}$, we consider it to be negligible in our consideration of "typical" elements of $\mathscr{B}$. Thus, for all intents and purposes, we can consider $\mathscr{B}$ to be identified with I.

In order to describe other features of our model of $\mathscr{B}$, we need some familiarity with the idea of Lebesgue measure. We will not yet attempt to define Lebesgue measure precisely, but we will describe some of the properties it should have. We ask the reader to believe that it exists until we examine it more rigorously.

A measure μ on a space X is a nonnegative function defined on a prescribed collection of subsets of X, the measurable sets. If A is a measurable set, the nonnegative number $\mu(A)$ is called the measure of A. Of course we will require that μ have certain properties so that it behaves as our intuition tells us a measure should behave. For example, we will require additivity: If A and B are measurable and disjoint, then $A \cup B$ is measurable and $\mu(A \cup B) = \mu(A) + \mu(B)$. This and other properties of measures will be discussed in §1.3.

The particular measure in which we are interested here is called Lebesgue measure (denoted μ_L) and is defined on certain subsets of the real line **R**. For the intervals

$$(a, b), \quad (a, b], \quad [a, b], \quad [a, b)$$

the Lebesgue measure is just the length, $b - a$. More generally, by the property of additivity, if

$$A = \bigcup_{i=1}^{n} A_i$$

is the finite disjoint union of finite intervals A_i, then A is Lebesgue measurable and

$$\mu_L(A) = \sum_{i=1}^{n} \mu_L(A_i)$$

Using the concept of Lebesgue measure, we can now formulate what we will call the *Borel principle*.

Borel principle. Suppose E is a probabilistic event occurring in certain Bernoulli sequences. Let $\mathcal{B}_E$ denote the subset of $\mathcal{B}$ for which the event occurs. Let B_E be the corresponding subset of I. Then the probability that E occurs, Prob(E), is equal to $\mu_L(B_E)$.

Let us show that this principle works for some simple probabilistic events.

1. E is the event that H appears on the first toss.

$$B_E = \{\omega \in I; \omega = .1\ldots\} = (\tfrac{1}{2}, 1]$$

so $\mu_L(B_E) = \tfrac{1}{2}$.

2. E is the event that the first N tosses are a prescribed sequence.

$$B_E = \{\omega = I; \omega = .a_1 a_2 a_3 \ldots a_N \ldots\}$$

where $a_1, \ldots, a_N$ are prescribed and everything else is arbitrary. Let $s = .a_1 a_2 \ldots a_N 00 \ldots$, then $B_E = (s, s + (1/2^N)]$ so that $\mu_L(B_E) = 1/2^N$ as expected.

3. E is the event that H occurs in the Nth place.

$$B_E = \{\omega \in I;\ \omega = .a_1\,a_2\ldots a_{N-1}\,1\,a_{N+1}\ldots\}$$

Fix a particular $s = .a_1\ldots a_{N-1}\,1000\ldots$. Then B_E contains the interval $(s, s + (1/2^N)]$. We can choose the $a_1,\ldots,a_{N-1}$ in 2^{N-1} different ways, and each of these intervals is disjoint from the others; so

$$\mu_L(B_E) = 2^{N-1}\left(\frac{1}{2^N}\right) = \frac{1}{2}$$

The shaded region is B_E for the event that H occurs on the third toss.

4. E is the event that, in the first N tosses, exactly k heads are seen.

$$B_E = \{\omega \in I;\ \omega = .a_1\,a_2\ldots a_N\ldots,\ \text{where } k \text{ of the first } N\ a_i\text{'s are } 1\}$$

Fix $.a_1,\ldots,a_N$, k of which are 1. Let $s = .a_1\,a_2\ldots a_N\,000\ldots$, so that B_E contains $(s, s + (1/2^N)]$. There are $\binom{N}{k}$ such intervals, all mutually disjoint, so

$$\mu_L(B_E) = \left(\frac{1}{2^N}\right)\binom{N}{k}$$

5. Start with X dollars and bet on a sequence of coin tosses. At each toss you win \$1.00 if a head shows up and you lose \$1.00 if a tail shows up. What is the probability that you lose all your original stake? To discuss this event we introduce some notation.

Rademacher Functions

For $\omega \in I$ we define the kth Rademacher function R_k by

(3) $$R_k(\omega) = 2a_k - 1$$

where $\omega = .a_1\,a_2\ldots$ is the binary expansion of ω. Note that

(4) $$R_k(\omega) = \begin{cases} +1 & \text{if } a_k = 1 \\ -1 & \text{if } a_k = 0 \end{cases}$$

so $R_k(\omega)$ represents the amount won or lost at the kth toss.

To familiarize ourselves with the Rademacher functions, we graph the first three, $R_1(\omega)$, $R_2(\omega)$, and $R_3(\omega)$.

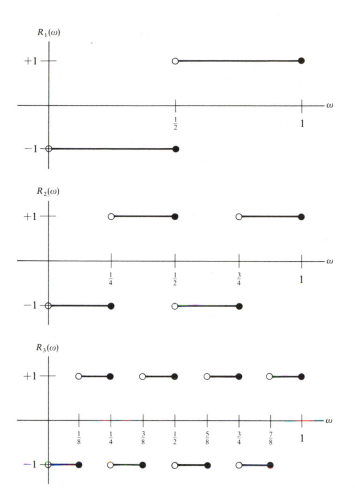

Using the R_k's, we can describe B_E for event 5. First consider the event E_k, representing loss of the original stake at the kth toss. Let

(5) $$S_k(\omega) = \sum_{l \leq k} R_l(\omega)$$

$S_k(\omega)$ gives the total amount won or lost at the kth stage of the game. Then

$$B_{E_k} = \{\omega \in I; S_l(\omega) > -X, l < k, \text{ and } S_k(\omega) = -X\}$$

and

$$B_E = \bigcup_{k=1}^{\infty} B_{E_k}$$

We will postpone the computation of $\mu_L(B_E)$ to §1.4 because B_E is not a finite union of intervals.

Now we return to the law of large numbers. Our assertion is that "roughly as many heads as tails turn up for a typical Bernoulli sequence." We formulate this statement mathematically as follows.

For $\omega \in I$, with $\omega = .a_1 \ldots a_N \ldots$, let $s_N(\omega) = a_1 + a_2 + \cdots + a_N$. This sum gives the number of heads in the first N terms of the Bernoulli sequence corresponding to ω. Now fix $\varepsilon > 0$ and consider

$$B_N = \left\{ \omega \in I; \left| \frac{s_N(\omega)}{N} - \frac{1}{2} \right| > \varepsilon \right\}$$

This set represents the event that, after the first N trials, there are *not* "roughly as many heads as tails." We can restate our assertion as follows.

Theorem 2.   (Weak law of large numbers)

$$\mu_L(B_N) \to 0 \quad \text{as} \quad N \to \infty$$

Proof.   We first describe B_N using Rademacher functions. Recall that

$$R_k(\omega) = 2a_k - 1$$

where $\omega = .a_1 a_2 \ldots a_k \ldots$. Thus

$$S_N(\omega) = \sum_{k=1}^{N} R_k(\omega) = 2(a_1 + a_2 + \cdots + a_N) - N = 2s_N(\omega) - N$$

Now

$$\left| \frac{s_N(\omega)}{N} - \frac{1}{2} \right| > \varepsilon \Leftrightarrow |2s_N(\omega) - N| > 2\varepsilon N$$

which is equivalent to $|S_N(\omega)| > 2\varepsilon N$. So, by altering ε slightly, we restate the theorem as follows.

$$\text{Let} \quad A_N = \{ \omega \in I; |S_N(\omega)| > N\varepsilon \}$$

$$\text{Then} \quad \mu_L(A_N) \to 0 \quad \text{as} \quad N \to \infty$$

To prove this form of the theorem, we will need the following special case of *Chebyshev's inequality*.

Lemma 3.   Let f be a nonnegative, piecewise constant function on $(0, 1]$. Let $\alpha > 0$ be given. Then

$$\mu_L(\{\omega \in I; f(\omega) > \alpha\}) < \frac{1}{\alpha} \int_0^1 f \, dx$$

(Here the integral $\int_0^1 f \, dx$ is the usual Riemann integral.)

Notice that we know how to compute $\mu_L(\{\omega \in I; f(\omega) > \alpha\})$ because $\{\omega \in I; f(\omega) > \alpha\}$ is a finite union of intervals.

Proof of lemma. When f is piecewise constant, there exist $x_1, \ldots, x_k$ with $0 = x_1 < \cdots < x_k = 1$ and $f = c_i$ on (x_i, x_{i+1}) $i = 1, \ldots, k-1$. (This is what we mean by piecewise constant.) Then

$$\int_0^1 f \, dx = \sum_{i=1}^{k-1} c_i(x_{i+1} - x_i) \geq \sum{}' c_i(x_{i+1} - x_i)$$

where Σ' means sum over the i's such that $c_i > \alpha$. Therefore,

$$\sum{}' c_i(x_{i+1} - x_i) > \alpha \sum{}'(x_{i+1} - x_i) = \alpha\mu_L(\{\omega \in I; f(\omega) > \alpha\})$$

so

$$\frac{1}{\alpha} \int_0^1 f \, dx > \mu_L(\{\omega \in I; f(\omega) > \alpha\}) \qquad\qquad \triangledown$$

Now we continue with the proof of our theorem. Notice that

$$A_N = \{\omega \in I; |S_N(\omega)| > N\varepsilon\}$$

$$= \{\omega \in I; |S_N(\omega)|^2 > N^2\varepsilon^2\}$$

An application of the preceding lemma gives

$$\mu_L(A_N) = \mu_L(\{\omega \in I; |S_N(\omega)|^2 > N^2\varepsilon^2\}) \leq \frac{1}{N^2\varepsilon^2} \int_0^1 S_N^2 \, dx$$

To exploit this inequality we need to compute $\int_0^1 S_N^2 \, dx$. However

$$\int_0^1 S_N^2 \, dx = \int_0^1 (\sum R_k)^2 \, dx = \sum_{k=1}^N \int_0^1 R_k^2 \, dx + \sum_{i \neq j} \int_0^1 R_i R_j \, dx$$

Because $R_k^2 = 1$, each of the first N terms is equal to one. What about

$$\int_0^1 R_i R_j \, dx \qquad i \neq j?$$

Suppose $i < j$. Let J be an interval of the form $(l/2^i, (l+1)/2^i]$, $0 \leq l < 2^i$. Then R_i is constant on J and R_j oscillates $2(j-i)$ times so that

$$\int_J R_j \, dx = 0$$

Thus

$$\int_0^1 R_i R_j \, dx = 0$$

which proves that

$$\int_0^1 S_N^2 \, dx = N$$

Thus

$$\mu_L(A_N) \le \left(\frac{1}{N^2 \varepsilon^2}\right) N = \frac{1}{N \varepsilon^2} \to 0 \quad \text{as} \quad N \to \infty \qquad \square$$

The astute reader has probably noticed that we haven't proved exactly what we said we intended to prove at the beginning of this section. Namely, we wanted to prove that, for a "typical" Bernoulli sequence,

(6) $$\frac{1}{2} - \frac{s_N(\omega)}{N} \to 0$$

By "typical" we should mean equation 6 fails on a set of zero probability. By the Borel principle, an event E has probability zero if the corresponding set $B_E \subset I$ has Lebesgue measure zero. The only sets we know thus far with zero Lebesgue measure are finite collections of points. When we extend Lebesgue measure to a collection of sets much bigger than the collection of intervals, we will find many more sets of measure zero. In fact we can describe these sets now without developing the general theory of Lebesgue measure.

Given a subset $A \subset \mathbf{R}$ and a countable collection of sets $\{A_i\}_{i=1}^\infty$, we will say the A_i's are a countable covering of A if $A \subset \bigcup_{i=1}^\infty A_i$.

Definition 4. A set $A \subset \mathbf{R}$ has Lebesgue measure zero if, for every $\varepsilon > 0$, there exists a countable covering $\{A_i\}$ of A by intervals such that

(7) $$\sum_{i=1}^\infty \mu_L(A_i) < \varepsilon$$

Remarks.

1. In this definition we can allow the A_i's to be finite unions of intervals.
2. If A has Lebesgue measure zero and $B \subset A$, then B has Lebesgue measure zero.

3. A single point has Lebesgue measure zero.
4. If $A_1, A_2, \ldots$ is a countable collection of sets, each having Lebesgue measure zero, then $\bigcup_{i=1}^{\infty} A_i$ has Lebesgue measure zero. In particular, countable sets have Lebesgue measure zero.

Proof. (Remarks 1, 2, and 3 are clear.) To prove remark 4, choose $\varepsilon > 0$. Because A_i has Lebesgue measure zero, there exists a countable collection of intervals $A_{i,1}, A_{i,2}, \ldots$ covering A_i such that

$$\sum_{j=1}^{\infty} \mu_L(A_{i,j}) < \frac{\varepsilon}{2^i}$$

The collection $\{A_{i,j}\}$ is countable, it covers $\bigcup_{i=1}^{\infty} A_i$, and

$$\sum_{i,j}^{\infty} \mu_L(A_{i,j}) = \sum_{i=1}^{\infty} \sum_{j=1}^{\infty} \mu_L(A_{i,j}) \leq \sum_{i=1}^{\infty} \frac{\varepsilon}{2^i} = \varepsilon \qquad \triangledown$$

Now let $N = \{\omega \in I; (s_n(\omega)/n) \to 1/2 \text{ as } n \to \infty\}$. N is called the set of *normal numbers*.

Theorem 5. (Strong law of large numbers) N^c has Lebesgue measure zero.

Remark. N^c is uncountable; in fact, N^c contains a "Cantor set."
Consider the map $\sigma : I \to I$ defined by

$$\sigma(\omega) = .a_1 \, 1 \, 1 \, a_2 \, 1 \, 1 \, a_3 \, 1 \, 1 \ldots$$

for $\omega = .a_1 a_2 a_3 \ldots$. This map is one to one, so its image is uncountable. Notice also that the image is contained in N^c. In fact, if $\omega' = \sigma(\omega)$, then $s_{3n}(\omega') \geq 2n$; so

$$\frac{s_{3n}(\omega')}{3n} \geq \frac{2}{3}$$

Now we will prove theorem 5. Let

$$A_n = \{\omega \in I; |S_n(\omega)| > \varepsilon n\} = \{\omega \in I; S_n^4(\omega) < \varepsilon^4 n^4\}$$

Then, by Chebyshev's inequality,

$$\mu_L(A_n) \leq \frac{1}{\varepsilon^4 n^4} \int_0^1 S_n^4 \, dx \qquad \text{where} \qquad \int_0^1 S_n^4 \, dx = \int_0^1 \left(\sum_{k=1}^{n} R_k\right)^4 dx$$

Multiplying out the integrand, we obtain five kinds of terms:

1. R_α^4 $\qquad\qquad$ $\alpha = 1, \ldots, n$
2. $R_\alpha^2 R_\beta^2$ $\qquad\quad$ $\alpha \neq \beta$

3. $R_\alpha^2 R_\beta R_\gamma$ $\alpha \neq \beta \neq \gamma$
4. $R_\alpha^3 R_\beta$ $\alpha \neq \beta$
5. $R_\alpha R_\beta R_\gamma R_\delta$ $\alpha \neq \beta \neq \gamma \neq \delta$

Because $R_\alpha^4 = 1$ and $R_\alpha^2 R_\beta^2 = 1$, $\int_0^1 R_\alpha^4 \, dx = \int_0^1 R_\alpha^2 R_\beta^2 \, dx = 1$.

We claim that the other terms all integrate to zero. In fact,

$$\int_0^1 R_\alpha^2 R_\beta R_\gamma \, dx = \int_0^1 R_\beta R_\gamma \, dx = 0$$

and

$$\int_0^1 R_\alpha^3 R_\beta \, dx = \int_0^1 R_\alpha R_\beta \, dx = 0$$

What about $R_\alpha R_\beta R_\gamma R_\delta$? Assume $\alpha < \beta < \gamma < \delta$ and consider an interval of the form $(l/2^\gamma, (l + 1)/2^\gamma]$. R_γ is constant on J and, because $\alpha < \beta < \gamma$, $R_\alpha R_\beta R_\gamma$ is constant on J as well. Finally, R_δ oscillates $2(\delta - \gamma)$ times on J, so

$$\int_J R_\alpha R_\beta R_\gamma R_\delta \, dx = 0$$

and

$$\int_0^1 R_\alpha R_\beta R_\gamma R_\delta \, dx = 0$$

Because there are n terms of the form R_α^4 and $3n(n - 1)$ terms of the form $R_\alpha^2 R_\beta^2$,

$$\int_0^1 S_n^4 \, dx = 3n^2 - 2n \leq 3n^2$$

and

$$\mu_L(A_n) \leq \left(\frac{1}{n^4 \varepsilon^4} \right) 3n^2 \leq \frac{3}{n^2 \varepsilon^4}$$

Lemma 6. Given $\delta > 0$, there exists a sequence $\varepsilon_1, \varepsilon_2, \ldots$ such that $\varepsilon_n \to 0$ and

$$\tag{8} \sum_{n=1}^\infty \frac{3}{n^2 \varepsilon_n^4} < \delta$$

Proof. Choose, for instance, ε_n such that

$$\varepsilon_n^4 = cn^{-1/2}$$

Then

$$\sum_{n=1}^{\infty} \frac{3}{\varepsilon_n^4 n^2} = \frac{3}{c} \sum_{n=1}^{\infty} \frac{1}{n^{3/2}}$$

If c is chosen large enough, this quantity is less than δ. $\triangledown$

Finally, for each n, set

$$B_n = \{\omega; |S_n(\omega)| > \varepsilon_n n\}$$

$\mu_L(B_n) < 3/\varepsilon_n^4 n^2$, hence $\sum_{n=1}^{\infty} \mu_L(B_n) < \delta$. Notice that the B_n's are finite unions of intervals since S_n is piecewise constant. Thus, if we can show that $N^c \subset \bigcup_{n=1}^{\infty} B_n$, the theorem will be proved.

Now $N^c \subset \bigcup_{n=1}^{\infty} B_n$ if $N \supset \bigcap_{n=1}^{\infty} B_n^c$. But, if $\omega \in \bigcap_{n=1}^{\infty} B_n^c$, then, for each n, $|S_n(\omega)| \leq \varepsilon_n n$; that is, $|S_n(\omega)/n| \leq \varepsilon_n$. Because $\varepsilon_n \to 0$, we conclude that $|S_n(\omega)/n| \to 0$; that is, $\omega \in N$. $\square$

Remarks.

1. We have just proven theorem 5 by showing that

$$(9) \qquad\qquad \mu_L(N^c) = 0$$

Notice that we needed a relatively sophisticated definition of "measure zero" to make sense of this statement, because N^c is such a bad set. In particular N^c is uncountable. (The only intervals of length zero are points, and N^c is not even a countable union of such sets.) Later, when we discuss the connection between measure and integration, we will see that this example provides a good illustration of why Riemann integration is inadequate for probability theory.

2. Notice that the strong law of large numbers (theorem 5) does not indicate at what point we can expect about as many heads as tails. In §3.8 we will discuss the central limit theorem, which has some bearing on this question.

Exercises for §1.1

1. Prove that the set $\mathcal{B}$ of Bernoulli sequences is uncountable by the Cantor diagonal argument.
2. **a.** Let $\omega \in I = (0, 1]$. Show that ω can be written in the form $\sum_{i=1}^{\infty} a_i/2^i$, $a_i = 0, 1$. Show that this expansion is unique when we restrict to nonterminating series.
   b. Show that, for any integer k, $\omega \in I$ can be written in the form $\sum_{i=1}^{\infty} a_i/k^i$, where $a_i = 0, 1, \ldots, k-1$. Show that the expansion is unique when we restrict to nonterminating series.
3. A gambler has an initial stake of one dollar. Calculate the probability of

ruin at times 1, 3, and 5. Show that the chance of eventual ruin is at least 70%.

4. Show that

$$\int R_{\gamma_1} R_{\gamma_2} \cdots R_{\gamma_n} \, dx = 0 \quad \text{or} \quad 1$$

for any sequence $\gamma_1 \leq \gamma_2 \leq \cdots \leq \gamma_n$. When is the answer one?

5. Define the Rademacher functions on the whole real line by requiring them to be periodic of period one—that is, by setting $R_k(x + 1) = R_k(x)$. With this definition, show that $R_{k+1}(x) = R_k(2x)$ and, by induction, that $R_k(x) = R_1(2^{k-1}x)$.

6. Show that

$$R_n(x) = -\text{sgn}[\sin(2\pi 2^{n-1} x)]$$

except at a finite number of points. (*Notation:* For any number a, sgn a is *one* if a is positive and *minus one* if a is zero or negative.) We will see later in the text that some interesting analogies exist between the Rademacher functions and the functions $\sin(2\pi 2^{n-1} x)$.

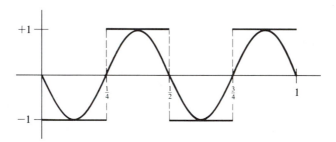

7. Prove that

$$2t - 1 = \sum_{k=1}^{\infty} R_k(t) 2^{-k}$$

8. Every number $\omega \in (0, 1]$ has a ternary expansion

$$\omega = \sum a_i 3^{-1}$$

with $a_i = 0$, 1, or 2 (see exercise 2). We can make this expansion unique by selecting, whenever ambiguity exists, the *nonterminating* expansion— that is, the expansion in which not all a_i's from a certain point on are equal to 0. With this convention, define

$$T_k(\omega) = a_k - 1$$

Draw the graph of T_k for $k = 1, 2, 3$. Can you discern a general pattern?

9. Obtain a recursion formula for the T_k's similar to the recursion formula for the R_k's in exercise 5.

10. Let C be the set of all numbers on the unit interval $[0, 1]$, which can be written in the form

$$\omega = \sum_{k=1}^{\infty} a_k 3^{-k}$$

with $a_k = 0$ or 2. Show that C is uncountable. (C is called the *Cantor set*.) (*Hint:* Use the Cantor diagonal process.)

11. Prove that the Cantor set (see exercise 10) can be constructed by the following procedure: From $[0, 1]$ remove the middle third, $(\frac{1}{3}, \frac{2}{3})$; from the remainder—that is, the intervals $[0, \frac{1}{3}]$ and $[\frac{2}{3}, 1]$—remove the middle thirds, and so on, ad infinitum. The remainder is the Cantor set.

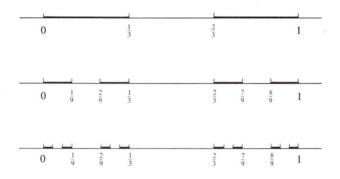

12. Show that the Cantor set is of measure zero.
13. Describe geometrically the set $\sigma(I)$ discussed on page 9.
14. Show that the nonnormal numbers are dense in the unit interval.
15. **a.** Show that a positive number c_3 exists such that, for all N,

$$\int [S_N(x)]^6 \, dx \le c_3 N^3$$

  b. Let A_n be the set $\{\omega \in I; |S_n(\omega)| > \varepsilon n\}$. Show that the Lebesgue measure of A_n is less than $c_3 \varepsilon^{-6} n^{-3}$.

16. More generally, show that a positive number c_K exists such that, for all N,

$$\int [S_N(x)]^{2K} \, dx \le c_K N^K$$

17. Prove a refinement of the strong law of large numbers, which says that

$$\frac{S_N}{N^{\delta}} \to 0 \quad \text{as} \quad N \to \infty$$

for any $\delta > \frac{1}{2}$. (*Hint:* Use exercise 16. We will see later that, for $\delta = \frac{1}{2}$, the situation is much more interesting.)

18. Prove that

$$\int_0^1 e^{tS_n(x)}\,dx = \left(\frac{e^t + e^{-t}}{2}\right)^n$$

(*Hint:* By induction. Write

$$\int_0^1 e^{tS_n(x)}\,dx = \int_0^1 e^{tS_{n-1}(x)}e^{tR_n(x)}\,dx$$

Break up the unit interval into 2^{n-1} equal subintervals on each of which R_{n-1} is constant. Show that

$$\int_J e^{tS_n(x)}\,dx = \left(\frac{e^t + e^{-t}}{2}\right)\int_J e^{tS_{n-1}(x)}\,dx$$

if J is one of these intervals.)

19. From exercise 18, derive the formula

$$\int_0^1 S_n(x)^{2K}\,dx = \left(\frac{d}{dt}\right)^{2K}\left.\left(\frac{e^t + e^{-t}}{2}\right)^n\right|_{t=0}$$

20. Let f be a nonnegative monotone function defined on the unit interval. Prove Chebyshev's inequality

$$\mu_L(\{\omega \in I; f(\omega) > \alpha\}) < \frac{1}{\alpha}\int_0^1 f\,dx$$

with the integral on the right being the Riemann integral.

21. We have already defined Lebesgue measure for two kinds of sets: finite unions of intervals and sets of Lebesgue measure zero. Show that these two definitions are not contradictory; that is, show that the interval $[a, b]$, $a < b$, is *not* a set of measure zero. (*Hint:* Use the Heine–Borel property of compact sets.)

§1.2 Randomness

In §1.1 we saw how to identify the set $\mathscr{B}$ of Bernoulli sequences with the set of points on the unit interval I. In terms of this identification, a probabilistic event E, associated with Bernoulli sequences, gets identified with a subset B_E of I. We saw that, at least for simple events, the Borel principle applies; that is,

(1) $$\mathrm{Prob}(E) = \mu_L(B_E)$$

We will attempt in this section to describe some slightly more complicated probabilistic events in measure theoretic terms.

Example 1. Gambler's Ruin

A gambler has X dollars and bets at even odds on a coin flip. What is the probability of his ruin?

We discussed this event already in §1.1. We showed that

$$B_E = \bigcup_{k=1}^{\infty} B_{E_k}$$

where

$$B_{E_k} = \{\omega \in I; S_l(\omega) > -X, l < k, \text{ and } S_k(\omega) = -X\}$$

After developing some measure theoretical tools, we will see that

(2) $$\mu_L(B_E) = \sum_{k=1}^{\infty} \mu(B_{E_k}) = 1$$

In other words, with probability one, if a gambler bets long enough, he will eventually lose all his money no matter how big his initial stake.

Example 2. Random Patterns

Pick a finite pattern of coin tosses, for example, T H H T. Let E be the event that T H H T occurs in a given Bernoulli sequence. Then

$$B_E = \{\omega \in I; \text{ there exists } n_0$$

$$\text{with } R_{n_0}(\omega) = -1, R_{n_0+1}(\omega) = 1, R_{n_0+2}(\omega) = 1, \text{ and } R_{n_0+3}(\omega) = -1\}$$

We will prove in §1.4 that this set is of measure one. In fact we will prove that, if one fixes *any* finite pattern, this pattern appears infinitely often in a Bernoulli sequence with probability one.

This result can be interpreted as follows. Put a monkey in front of a telegraph key and let him punch a series of dots and dashes as he pleases. With probability one, the monkey will eventually tap out in Morse code all the sonnets of Shakespeare *infinitely often*.

Example 3. Random Variables

In example 1 let R_n be the amount of money won or lost at the nth toss. R_n can be thought of as a function on the set $\mathscr{B}$ or, via the identification $\mathscr{B} \leftrightarrow I$, as a function on the unit interval. It is, of course, just the nth Rademacher function, discussed in §1.1. R_n is a typical example of what probabilists call a *random variable*. It is a *variable*—that is, a quantity that one can measure each time one performs a sequence of Bernoulli trials—and it is *random*, because the values it assumes are a matter of hazard or chance. Another example of a random variable is the sum

$$S_n = \sum_{k=1}^{n} R_k$$

which is the total amount won or lost by the nth stage of the game. Notice that the set B_{E_k} in example 1 is completely described by the S_n's. This is not surprising. Most interesting random events are describable by random variables. For instance, consider *winning streaks*. Suppose that, starting at time $t = n$, a gambler tosses an unbroken sequence of heads for a certain length of time. The relevant random variable connected with this phenomenon is the variable l_n, which counts the number of times H occurs consecutively starting with the nth toss.

Example 4. Expectation Values

Let $\mathscr{B}$ be the set of Bernoulli sequences. A random variable associated with the Bernoulli process is, by definition, a function $f : \mathscr{B} \to \mathbf{R}$. Thanks to the identification of $\mathscr{B}$ with I, we can also think of f as a function on I. In Chapter 2 we will address the question of what kinds of functions correspond to the "physically interesting" random variables. For these functions we will be able to define the Lebesgue integral

(3) $$\int_I f \, d\mu_L$$

The probabilists call equation 3 the *expectation value* of the random variable f. Roughly speaking, it is the value that f is "most likely" to assume in a series of frequently repeated experiments. To use a simplistic example, for the Rademacher function R_n, the integral in equation 3 turns out to be just the usual Riemann integral

$$\int_0^1 R_n \, dx$$

which, as we saw in the previous section, is zero; that is, the "most likely" value of R_n is zero (even though R_n takes only the values $+1$ and -1). We will justify this somewhat paradoxical assertion in §2.6.

Example 5. Random Walks

A Bernoulli sequence, that is, a sequence of coin tosses, can be considered to describe a random walk on the real line. That is, a particle is placed at the origin; a flip of a head causes the particle to move one step forward, and a tail moves it one step backward. As one tosses the coin an infinite number of times, the particle moves erratically backward and forward along the real line. We will call the path traced out by such a particle a *random path* and the sequence of motions itself a *random walk*. Obviously each Bernoulli sequence gives rise to a random path and vice versa. If we denote by $\mathscr{R}$ the set of random paths, we can identify $\mathscr{R}$ with $\mathscr{B}$ and, by means of binary expansions, both $\mathscr{R}$ and $\mathscr{B}$ with the unit interval I. Probabilistic events associated with $\mathscr{R}$ can be reinterpreted as events associated with $\mathscr{B}$ and vice versa. For example

$$\text{gambler's ruin} \leftrightarrow \text{passing through } -X \text{ for the first time}$$

In probability jargon the space of all possible outcomes of a probabilistic process is called the *sample space*. For Bernoulli sequences, the sample space is $\mathscr{B}$; for random walks the sample space is $\mathscr{R}$. For all intents and purposes, $\mathscr{R}$ and $\mathscr{B}$ are identical, even though one thinks of $\mathscr{R}$ in connection with the motion of particles and $\mathscr{B}$ in connection with games of chance.

Example 6. Random Walks with Pauses

To perform a random walk with pauses, one needs a gadget of the type depicted in the figure below. Place a particle at the origin of the real line and spin the pointer. If it lands on $+1$, move the particle one unit to the right; if it lands on -1, move the particle one unit to the left; and, if it lands on 0,

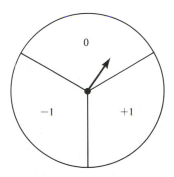

leave the particle fixed. By repeating this operation infinitely often, we get a *random walk with pauses*. Let $\mathscr{B}_p$ be the sample space of this process. Identify $\mathscr{B}_p$ with I using ternary expansions of points, $\omega \in I$; that is, each $\omega \in I$ can be written as

(4)
$$\omega = \sum_{k=1}^{\infty} \frac{a_k}{3^k} \qquad a_k = 0, 1, 2$$

The ternary expansion of ω is then

$$\omega = .a_1 a_2 a_3 \dots$$

Notice that $\frac{1}{3} = .1000\dots$ or $.0222\dots$; so, in order to make ternary expansions unique, we will always choose the nonterminating expansion in cases like the one above (see §1.1, exercise 2). Now make the identification

$$+1 \leftrightarrow 1$$

$$0 \leftrightarrow 0$$

$$-1 \leftrightarrow 2$$

to identify a random walk with pauses with the digits in such a ternary expansion. This identification gives a map $I \to \mathscr{B}_p$. The Borel principle in this instance says that, if an event E associated with this random process corresponds to the subset B_E of I, then, just as before,

(5)
$$\text{Prob}(E) = \mu_L(B_E)$$

We suggest you check equation 5 for a few simple events. (See exercise 4.)

Example 7. Random Walks in the Plane

For two-dimensional random walks, we need a gadget similar to the one we used on the previous page:

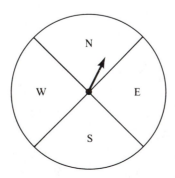

Let $\mathbf{Z}^2 = \{(m, n); m, n \text{ integers}\}$, which is the *integer lattice* in the plane. Place a particle at $(0, 0) \in \mathbf{Z}^2$ and spin the pointer. If it lands on N, move the particle to $(0, 1) \in \mathbf{Z}^2$; if it lands on E, move the particle to $(1, 0) \in \mathbf{Z}^2$, and so on. By repeating this operation ad infinitum, one produces a random walk, the successive stages of which are indexed by an infinite sequence such as

(6) $$\text{N S S E W E} \ldots$$

Let $\mathscr{B}_{\text{plane}}$ be the sample space of this process—that is, the set of all sequences like the one in display 6. We can identify each sequence with a point $\omega \in I$ using base-four expansions; that is, $\omega \in I$ can be written as

(7) $$\omega = \sum_{k=1}^{\infty} \frac{a_k}{4^k} \qquad a_k = 0, 1, 2, 3$$

The base-four expansion of ω is then $.a_1 a_2 a_3 \ldots$, which can be identified with a sequence like the one in display 6 by means of the correspondence

$$0 \leftrightarrow \text{East}$$

$$1 \leftrightarrow \text{West}$$

$$2 \leftrightarrow \text{North}$$

$$3 \leftrightarrow \text{South}$$

Of course we must deal with the problem of nonuniqueness in this identification as above, by selecting nonterminating rather than terminating expansions whenever ambiguity exists. Just as in example 6, to every event E associated with this process there corresponds a subset B_E of I. We urge you to check that

$$\text{Prob}(E) = \mu_L(B_E)$$

for a few simple, typical events. (See exercise 5.)

Example 8. The Discrete Dirichlet Problem

Let Ω be a smooth, bounded region in the plane with boundary B. An important problem in electrostatics is the *Dirichlet problem:* Given a continuous function f on B, find a function u satisfying

(8)
$$\Delta u = 0 \qquad \text{in } \Omega$$
$$u = f \qquad \text{on } B$$

where $\Delta u = (\partial^2 / \partial^2 x) u + (\partial^2 / \partial^2 y) u$.

This problem has a discrete analogue that is itself quite interesting. Let Ω

be a finite subset of $\mathbf{Z}^2$. A point $p = (m, n)$ of Ω is an *interior point* if its four next-door neighbors

$$(m, n+1), \quad (m+1, n), \quad (m, n-1), \quad \text{and} \quad (m-1, n)$$

are also in Ω; otherwise, p is a *boundary point*. For instance, in the figure below, p_1 is an interior point and p_2 a boundary point of the shaded region.

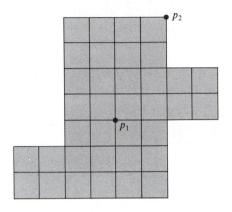

For a function u on $\mathbf{Z}^2$, we define $\Delta_{\text{discrete}} u$ by the formula

$$
(9) \qquad
\begin{aligned}
&(\Delta_{\text{discrete}} u)(m, n) \\
&= \frac{u(m, n+1) + u(m, n-1) + u(m+1, n) + u(m-1, n)}{4} - u(m, n)
\end{aligned}
$$

[Notice that the first term on the right is just the average of u over the next-door neighbors of the point (m, n).] The discrete analogue of the Dirichlet problem is to find a function $u : \mathbf{Z}^2 \to \mathbf{R}$ such that

$$(10) \qquad\qquad\qquad \Delta_{\text{discrete}} u = 0$$

at the interior points of Ω, and

$$(11) \qquad\qquad\qquad u = f$$

on the boundary, $\partial\Omega$, of Ω, f being a given function on $\partial\Omega$. One can solve this problem elegantly by using the random walk described in example 7: Given a point $p \in \Omega$ and a random path ω starting at p, let $F(\omega, p)$ be the value of f at the first point at which ω hits $\partial\Omega$. [If ω never hits the boundary, set $F(\omega, p) = 0$.] If we fix p and regard F as a function of the random path ω alone, then F is a *random variable* in the sense of example 3. We will show in §2.8 that its expectation value is the value at p of the solution of the Dirichlet problem described in equations 10 and 11.

Example 9. Randomized Series

Probabilistic considerations have another way of entering into classical analysis. Consider the series

$$\sum_{n=1}^{\infty} \frac{1}{n} \quad \text{and} \quad \sum_{n=1}^{\infty} \frac{(-1)^n}{n}$$

The first of these series diverges, whereas the second converges. We can enrich this problem by adding a probabilistic component. Consider a general series

$$\sum_{n=1}^{\infty} \frac{\pm 1}{n}$$

where the plus or minus is determined by the flip of a coin; that is, for each $\omega \in I$ we get a series

(12) $$\sum_{n=1}^{\infty} R_n(\omega)\left(\frac{1}{n}\right)$$

with R_n being the nth Rademacher function. Now let E be the event that this series converges:

$$B_E = \left\{ \omega \in I; \sum_{n=1}^{\infty} R_n(\omega)\left(\frac{1}{n}\right) \text{converges} \right\}$$

What is $\mu_L(B_E)$?

In §3.3 we give a series of exercises in which we sketch a proof that $\mu_L(B_E) = 1$. The intuition behind this result is that a typical Bernoulli sequence has roughly as many pluses (heads) as minuses (tails).

Example 10

We end this section by considering a collection of sample spaces that includes all of those we have considered up to this point.

Take n marbles of k various colors. Say the colors are labeled $c_1, c_2, \ldots, c_k$ and suppose that, of the N marbles, N_j of them have the color c_j, $1 \le j \le k$. Now put all of the marbles into a cylindrical wire cage that can be spun on its axis to mix the marbles fairly well within the cage. After the marbles are mixed, a blindfolded assistant removes one marble from the cage. If the marble has the color c_j, one gets as a reward a preassigned number, r_j, of dollars. (Incidentally, we will allow r_j to be positive *or* negative.) After the color of the marble is recorded, the marble is returned to the cage and the process is repeated.

The probability that the color c_j will be chosen is

$$p_j = \frac{N_j}{N}$$

and

(13) $$\sum_{j=1}^{k} p_j = 1 \qquad \text{because} \qquad \sum_{j=1}^{k} N_j = N$$

Notice that this game can serve as a model for coin tossing (and random walks) by allowing only two colors of equal number, say c_1 = red and c_2 = white, and setting the rewards at $+\$1.00$ for a red marble and $-\$1.00$ for a white marble.

With three colors of equal number, say c_1 = blue, c_2 = white, and c_3 = red, and rewards $r_1 = +\$1.00$, $r_2 = 0$, and $r_3 = -\$1.00$, we get a model for the random walk with pauses.

If we alter our process slightly by allowing the r_i's to be vectors in $\mathbf{R}^2$, we can model random walks in the plane; namely, take four colors of equal numbers with $r_1 = (0, 1)$, $r_2 = (1, 0)$, $r_3 = (-1, 0)$, and $r_4 = (0, -1)$.

In §2.6 we will develop a measure theoretic model for this process based on a "Borel principle" similar to that in the preceding examples.

Exercises for §1.2

1. Under the correspondence $\mathscr{B} \leftrightarrow I$, describe the subset of I corresponding to the event that a run of 15 heads will occur before a run of 11 tails.
2. Describe the subset of I corresponding to the event that no run of heads longer than 15 occurs in a Bernoulli sequence.
3. Prove that the pattern HT has to occur infinitely often in a Bernoulli sequence (with probability one) using the Borel principle.
4. With the ternary numbers as a model for the random walk with pauses, test the Borel principle by using it to compute the probability of
   a. a pause at time $t = 1$.
   b. a pause at time $t = n$.
   c. forward motion at times $t = 1, 2, 3, \ldots, n$.
   d. forward motion at times $t = k, k + 1, \ldots, k + n$.
5. With the quaternary numbers as a model for the random walk in the plane, test the Borel principle by using it to compute the probability that
   a. the first move is due east.
   b. the nth move is due east.
   c. the first n moves lie on a straight line.
6. With the ternary numbers as a model for the random walk with pauses, prove that with probability one an infinite number of pauses occur. (*Hint:* See §1.1, exercise 12.)

7. Sum the series

$$\sum \pm \frac{1}{2^n}$$

by the following procedure. For each Bernoulli sequence, put a $(+)$ sign in the kth place if a head comes up and a $(-)$ sign if a tail comes up. What is the sum? (*Hint:* See §1.1, exercise 7.)

8. Let Ω be the subset

$$\{(0,0), (1,0), (0,1), (-1,0), (0,-1)\}$$

of $\mathbf{Z}^2$. (That is, Ω consists of the origin and its four next-door neighbors.) Check directly that the recipe described in example 8 for solving the "discrete Dirichlet problem" on Ω is correct.

9. For the process described in example 10, show that, if one uses an equal number of marbles of each color, the sample space of the process can be identified with the unit interval using expansions in base k.

10. For the ordinary random walk starting at the origin, show that the probability of a particle's being in position k at time $t = n$ is

$$(*)$$

$$0 \qquad \text{if } |k| > n \text{ or if } n + k \text{ is odd}$$

$$\left(\frac{1}{2^n}\right)\binom{n}{r}, \quad \text{where } r = \frac{n+k}{2} \qquad \text{otherwise}$$

11. (On Markov processes.) Let $P = (P_{ij})$, $-\infty < i, j < \infty$, be an infinite matrix with the following properties:

$$(**)$$

  (i) $P_{ij} \geq 0$
  (ii) $\sum_j P_{ij} = 1$ for all i
  (iii) For fixed i, $P_{ij} = 0$ for all but finitely many j's.

For the "generalized random walk" associated with P, a particle moves along the line according to the following probabilistic rule: If the particle is at position i at time $t = n$, then at time $t = n + 1$ it can be at any position j for which $P_{ij} \neq 0$, and the probability of its being there is P_{ij}. (For instance, if $P_{ii} = 1$ and $P_{ij} = 0$ for $i \neq j$, then the particle stays forever at its initial position.) The matrix P is called the matrix of transition probabilities associated with the process.

a. Show that the process described in example 10 is a process of this kind. (Think of the position of the particle as being the total number of dollars won or lost by time $t = k$.)

b. For the process described in example 10, show that the matrix of transition probabilities is of the form $P_{ij} = P_{i-j}$.

c. Show, conversely, that if $P_{ij} = P_{i-j}$ the corresponding process is a process of the kind described in example 10.

12. a. Show that, if P and Q are matrices of the form in equation (**), the usual matrix product PQ is well-defined and is of the same form.

   b. Show that, for the generalized random walk associated with P, if the position of a particle at time zero is i, the probability that its position at time $t = n$ is j is just the $i - j$th entry of the matrix P^n.

13. Show that, for the matrix $P_{i,i+1} = P_{i+1,i} = \frac{1}{2}, P_{i,j} = 0$; otherwise, the generalized random walk is the usual random walk. Derive the formula (*) of exercise 10 by computing directly the $i - j$th entry of P^n. (*Hint:* Consider the vector space V consisting of all finite sums:

$$\sum c_k e^{kt} \qquad c_k \in \mathbf{R}$$

On this vector space consider the linear mapping "multiplication by $(e^{-t} + e^t)/2$." Show that, if we take for a basis of V

$$\dots e^{-kt}, \dots e^{-t}, 1, e^t, \dots, e^{kt}, \dots$$

then, in terms of this basis, this linear mapping has P as its matrix.)

14. Can you construct a measure theoretic model for random walks in space similar to the measure theoretic model for random walks in the plane? (*Hint:* Expansions in base six.)

§1.3 Measure Theory

We mentioned earlier that Lebesgue measure assigns to each set A, belonging to a certain collection of subsets of $\mathbf{R}$, a nonnegative number $\mu_L(A)$ called the Lebesgue measure of A. We also mentioned that μ_L has certain additivity properties. We will now study these properties in more detail. We need to begin with a large number of technical definitions. Keep in mind the vague notion of Lebesgue measure we have already discussed so as to put these technicalities in perspective.

Let X be a fixed set. Suppose A and B are subsets of X. We recall the following notation:

Notation	*Meaning*
$\varnothing$	empty set
$A \cup B = \{x \in X; x \in A \text{ or } x \in B\}$	union of A and B
$A \cap B = \{x \in X; x \in A \text{ and } x \in B\}$	intersection of A and B
$A^c = \{x \in X; x \notin A\}$	complement of A
$B - A = \{x \in X; x \in B \text{ and } x \notin A\}$	B minus A
$S(A, B) = (A - B) \cup (B - A)$	symmetric difference of A and B (see figure, page 25)

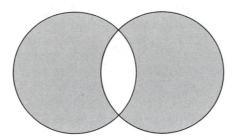

A *ring* of sets in X is a collection $\mathscr{R}$ of subsets of X satisfying the following two properties

1. $A \cup B \in \mathscr{R}$ whenever $A, B \in \mathscr{R}$
2. $A - B \in \mathscr{R}$ whenever $A, B \in \mathscr{R}$

Remark. $\varnothing \in \mathscr{R}$ since $A - A = \varnothing$.

Two examples with which we will soon be very familiar follow.

Example 1.   Let 2^X denote the set of all subsets of X; 2^X is a ring.

Example 2.   Let $X = \mathbf{R}^n$. Suppose $(a_1, \ldots, a_n)$ and $(b_1, \ldots, b_n)$ are given, with each $a_i \le b_i$, $i = 1, 2, \ldots, n$.

Let A be the set of points $x \in \mathbf{R}^n$ such that

(1) $$a_i \le x_i \le b_i \qquad i = 1, \ldots, n$$

A is called a multi-interval. More generally, a *multi-interval* is a set of the form shown with perhaps some of the $\le$'s replaced by a $<$.

Define $\mathscr{R}_{\text{Leb}}$ by $A \in \mathscr{R}_{\text{Leb}} \Leftrightarrow A = \bigcup_{i=1}^{N} A_i$, where the A_i's are a disjoint collection of multi-intervals. We let the reader check that $\mathscr{R}_{\text{Leb}}$ is a ring.

Now, fix a ring $\mathscr{R}$ of subsets of X. Let μ be a nonnegative set function on $\mathscr{R}$; that is, to each $A \in \mathscr{R}$, μ assigns a nonnegative number $\mu(A)$.

Definition 3. μ is *additive* if $\mu(A \cup B) = \mu(A) + \mu(B)$ whenever $A, B \in \mathscr{R}$ are disjoint.

Example 4. $\mathscr{R} = \mathscr{R}_{\text{Leb}}$. Suppose $A \in \mathscr{R}_{\text{Leb}}$ is a multi-interval described by the inequalities

$$a_i \le x_i \le b_i \qquad i = 1, \ldots, n$$

(Again, some $\le$'s may be replaced by $<$'s.) We define

(2) $$\mu(A) = (b_1 - a_1)(b_2 - a_2) \cdots (b_n - a_n)$$

More generally, if $A = \bigcup_{i=1}^{N} A_i$ is a disjoint union of multi-intervals, we define

(3)
$$\mu(A) = \sum_{i=1}^{N} \mu(A_i)$$

Then μ is a well-defined additive set function on $\mathscr{R}_{\text{Leb}}$.

Proposition 5. Let $\mathscr{R}$ be a ring of subsets of X and μ an additive, nonnegative set function on $\mathscr{R}$. Then

1. $\mu(\varnothing) = 0$.
2. (monotonicity) If $A, B \in \mathscr{R}$ with $A \subseteq B$, then $\mu(A) \leq \mu(B)$.
3. (finite additivity) If $A_1, A_2, \ldots, A_n \in \mathscr{R}$ are mutually disjoint, then $\mu(\bigcup_{i=1}^{n} A_i) = \sum_{i=1}^{n} \mu(A_i)$.
4. (lattice property) If $A, B \in \mathscr{R}$ then $\mu(A \cup B) + \mu(A \cap B) = \mu(A) + \mu(B)$.
5. (finite subadditivity) For any $A_1, \ldots, A_n \in \mathscr{R}$, $\mu(\bigcup_{i=1}^{n} A_i) \leq \sum_{i=1}^{n} \mu(A_i)$.

Proof.

1. $A \in \mathscr{R}$, $\mu(A) = \mu(A \cup \varnothing) = \mu(A) + \mu(\varnothing)$,   so $\mu(\varnothing) = 0$.
2. $B = (B - A) \cup A$ is disjoint, so $\mu(B) = \mu(B - A) + \mu(A) \geq \mu(A)$.
3. Induction on n.
4.
$$A = (A - B) \cup (A \cap B)$$
$$B = (A - B) \cup (A \cap B)$$
$$A \cup B = (A - B) \cup (B - A) \cup (A \cap B) .$$
so
$$\mu(A) = \mu(A - B) + \mu(A \cap B)$$
$$\mu(B) = \mu(B - A) + \mu(A \cap B)$$
$$\mu(A \cup B) = \mu(A - B) + \mu(B - A) + \mu(A \cap B)$$
$$= \mu(A) + \mu(B) - \mu(A \cap B)$$
5. Induction on n; case $n = 2$ follows from item 4. $\square$

So far we have done nothing very deep. We have just given an abstract setting for the situation in example 4. Our eventual purpose is to extend the definition of the set function in example 4 to a much larger ring of subsets of **R**. For instance, this ring should contain the sets of measure zero described in §1.1. In order to carry out this extension in a natural way, we will need the following refinement of additivity. As the proof of theorem 7 will suggest, this property is much more intricate than finite additivity.

Definition 6.

1. Let $\mathscr{R}$ be a ring of subsets of X and μ an additive set function on $\mathscr{R}$. We say μ is *countably additive* on $\mathscr{R}$ if, given any countable collection $\{A_i\}_{i=1}^{\infty} \subset \mathscr{R}$ with the A_i's mutually disjoint *and* such that $A = \bigcup_{i=1}^{\infty} A_i$ is also in $\mathscr{R}$, then

(4)
$$\mu(A) = \sum_{i=1}^{\infty} \mu(A_i)$$

2. A countably additive, negative set function μ on a ring $\mathcal{R}$ in X is called a *measure*.

Warning. Equation 4 makes no sense unless we assume $A \in \mathcal{R}$.

Theorem 7. If $X = \mathbf{R}^n$, $\mathcal{R} = \mathcal{R}_{\text{Leb}}$, and μ is the set function in example 4, then μ is a measure.

Lemma 8. Let $A \in \mathcal{R}_{\text{Leb}}$, and let $\varepsilon > 0$ be given. There exist $F, G \in \mathcal{R}_{\text{Leb}}$ such that F is closed, G is open, $F \subseteq A \subseteq G$, and

$$\mu(F) \geq \mu(A) - \varepsilon$$

$$\mu(G) \leq \mu(A) + \varepsilon$$

Proof. Suppose A is a multi-interval given by the inequalities

$$a_i \leq x_i \leq b_i \qquad i = 1, \ldots, n$$

where some of the $\leq$'s may be replaced by $<$'s. We can find a δ such that

$$\prod_{i=1}^{n} [(b_i - \delta) - (a_i + \delta)] = \prod_{i=1}^{n} (b_i - a_i - 2\delta) \geq \prod_{i=1}^{n} (b_i - a_i) - \varepsilon$$

and

$$\prod_{i=1}^{n} [(b_i + \delta) - (a_i - \delta)] = \prod_{i=1}^{n} (b_i - a_i + 2\delta) \leq \prod_{i=1}^{n} (b_i - a_i) + \varepsilon$$

Let F be given by the inequalities

$$a_i + \delta \leq x_i \leq b_i - \delta \qquad i = 1, \ldots, n$$

and G by the inequalities

$$a_i - \delta < x_i < b_i + \delta \qquad i = 1, \ldots, n$$

We then have

$$\mu(F) \geq \mu(A) - \varepsilon$$

$$\mu(G) \leq \mu(A) + \varepsilon$$

Now, if $A = \bigcup_{i=1}^{k} A_i$ is a disjoint union of multi-intervals, find for each A_i an F_i and G_i such that

$$\mu(F_i) \ge \mu(A_i) - \frac{\varepsilon}{k}$$

$$\mu(G_i) \le \mu(A_i) + \frac{\varepsilon}{k}$$

Then, with $F = \bigcup_{i=1}^{k} F_i$ and $G = \bigcup_{i=1}^{k} G_i$, we have

$$\mu(F) = \sum_{i=1}^{k} \mu(F_i) \ge \sum_{i=1}^{k} \left[\mu(A_i) - \frac{\varepsilon}{k} \right] = \mu(A) - \varepsilon$$

$$\mu(G) \le \sum_{i=1}^{k} \mu(G_i) \le \sum_{i=1}^{k} \left[\mu(A_i) + \frac{\varepsilon}{k} \right] = \mu(A) + \varepsilon \qquad \triangledown$$

Now, take $\{A_i\}_{i=1}^{\infty}$ to be a disjoint collection of sets in $\mathscr{R}_{\text{Leb}}$, and suppose $A = \bigcup_{i=1}^{\infty} A_i$ is also in $\mathscr{R}_{\text{Leb}}$. Notice that $\bigcup_{i=1}^{N} A_i \subset A$, so

(5) $$\mu(A) \ge \mu\left(\bigcup_{i=1}^{N} A_i \right) = \sum_{i=1}^{N} \mu(A_i) \qquad \text{for every } N$$

Thus

(6) $$\mu(A) \ge \sum_{i=1}^{\infty} \mu(A_i)$$

Choose a closed set $F \subseteq A$ such that $\mu(F) \ge \mu(A) - \varepsilon$, and for each A_i choose an open set G_i containing A_i with $\mu(G_i) \le \mu(A_i) + \varepsilon/2^i$.

Because F is closed and bounded, it is compact. Because it is covered by the G_i's, it must be covered by a finite number of them, say $G_1, G_2, \ldots, G_N$. Then

$$\mu(A) - \varepsilon \le \mu(F) \le \mu\left(\bigcup_{i=1}^{N} G_i \right) \le \sum_{i=1}^{N} \mu(G_i) \le \sum_{i=1}^{N} \left[\mu(A_i) + \frac{\varepsilon}{2^i} \right] \le \sum_{i=1}^{\infty} \mu(A_i) + \varepsilon$$

Being true for all ε, this yields

(7) $$\mu(A) \le \sum_{i=1}^{\infty} \mu(A_i)$$

Putting inequality 7 together with inequality 6 shows that μ is a measure. $\square$

We have now constructed a measure on a collection of subsets of $\mathbf{R}^n$. The sets on which this measure is defined, $\mathscr{R}_{\text{Leb}}$, are very simple, however. As remarked above, the property of countable additivity will allow us to extend this measure to a much larger ring of sets.

Let μ be a measure on a ring $\mathscr{R}$ in X. We attempt to extend μ to the ring 2^X by mimicking the definition of measure zero in §1.1.

Definition 9. Let A be a subset of X. A number $l \geq 0$ will be called an *approximate outer measure* of A if there exists a covering of A by a countable collection of sets $A_1, A_2, A_3, \ldots$ with each $A_i \in \mathcal{R}$ such that

$$(8) \qquad \sum_{i=1}^{\infty} \mu(A_i) \leq l$$

Remark. l is allowed to be $+\infty$.

Definition 10. Let A be a subset of X. The *outer measure* of A, $\mu^*(A)$, is the greatest lower bound of the set $\{l: l \text{ is an approximate outer measure of } A\}$. If this set is empty, then $\mu^*(A) = +\infty$.

We now have a set function, μ^*, on the ring 2^X. Unfortunately, μ^* is not generally a measure (see, for example, exercise 1). We will show, however, that μ^* is a measure on a large ring of subsets of X; this ring will be called the ring of measurable sets in X.

Proposition 11.

1. If $A \in \mathcal{R}$, then $\mu^*(A) = \mu(A)$.
2. If $A \subseteq B$, then $\mu^*(A) \leq \mu^*(B)$.
3. μ^* is countably subadditive; that is, if $A_1, A_2, A_3, \ldots$ are subsets of X, then $\mu^*(\bigcup_{i=1}^{\infty} A_i) \leq \sum_{i=1}^{\infty} \mu^*(A_i)$.

Proof.

1. Covering A by the sequence $A_1 = A, A_2 = \emptyset, A_3 = \emptyset, \ldots$, we see that $\mu(A)$ is an approximate outer measure for A, so

$$(9) \qquad \mu^*(A) \leq \mu(A)$$

To prove the other inequality, let $\varepsilon > 0$ be given. Because $\mu^*(A)$ is the greatest lower bound of all approximate outer measures of A, a cover $\{A_i\}_{i=1}^{\infty} \subset \mathcal{R}$ must exist such that

$$(10) \qquad \mu^*(A) + \varepsilon \geq \sum_{i=1}^{\infty} \mu(A_i)$$

Let $A_1' = A_1$, $A_2' = A_2 - A_1$, $A_3' = A_3 - (A_1 \cup A_2)$, and so on. Then the A_i''s are mutually disjoint and

$$(11) \qquad \mu^*(A) + \varepsilon \geq \sum_{i=1}^{\infty} \mu(A_i')$$

If we let $A_i'' = A_i' \cap A$, we have that $A_i'' \subseteq A$ for all i, the A_i'''s are mutually disjoint, and still

(12) $$\mu^*(A) + \varepsilon \geq \sum_{i=1}^{\infty} \mu(A_i'')$$

Now, since $A_i'' \subseteq A$ for all i and $\bigcup_{i=1}^{\infty} A_i \supset A$, we must have $\bigcup_{i=1}^{\infty} A_i'' = A$. So $\mu(A) = \sum_{i=1}^{\infty} \mu(A_i'')$ and thus

(13) $$\mu^*(A) + \varepsilon \geq \mu(A)$$

Because this inequality is true for all ε, we have

(14) $$\mu^*(A) \geq \mu(A)$$

  2. If l is an approximate outer measure for B, then surely it is for A. Thus

$$\mu^*(A) \leq \mu^*(B)$$

  3. Given $\varepsilon > 0$, for each i we can find a cover, $\{A_{i,j}\}_{j=1}^{\infty} \subset \mathcal{R}$, of A_i such that

(15) $$\mu^*(A_i) + \frac{\varepsilon}{2^i} \geq \sum_{j=1}^{\infty} \mu(A_{i,j})$$

Then the countable collection $\{A_{i,j}\}_{i,j=1}^{\infty}$ covers $A = \bigcup_{i=1}^{\infty} A_i$ so that

(16)
$$\mu^*(A) \leq \sum_{i,j=1}^{\infty} \mu(A_{i,j}) = \sum_{i=1}^{\infty} \sum_{j=1}^{\infty} \mu(A_{i,j})$$
$$\leq \sum_{i=1}^{\infty} \left[\mu^*(A_i) + \frac{\varepsilon}{2^i} \right]$$
$$= \varepsilon + \sum_{i=1}^{\infty} \mu^*(A_i)$$

This holds for all $\varepsilon > 0$, so

(17) $$\mu^*(A) \leq \sum_{i=1}^{\infty} \mu^*(A_i) \qquad \qquad \square$$

Remark.   This proof is essentially the same as that used in §1.1 to show that a countable union of sets of measure zero is itself of measure zero.

  Now our original ring, $\mathcal{R}$, is a subset of 2^X. We wish to find a larger ring, $\mathcal{M}$, containing $\mathcal{R}$, that will be the measurable sets. Our strategy will be to think of 2^X as a metric space and define a distance function on it, so that, roughly speaking, $\mathcal{M}$ will be the closure of $\mathcal{R}$ in 2^X with respect to this distance function. (For a quick review of metric spaces, see Appendix A.)

For $A, B \subset X$ we define the distance from A to B by

(18) $$d(A, B) = \mu^*[S(A, B)]$$

where $S(A, B)$ is the symmetric difference $S(A, B) = (A - B) \cup (B - A)$.

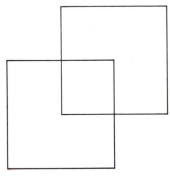

If *A* and *B* are the unit squares pictured, then $d(A, B) = 1\frac{1}{2}$.

Caution.

1. $d(A, B)$ may be $+\infty$.
2. Although we are calling d a distance function, $d(A, B) = 0$ does not necessarily imply $A = B$.

Proposition 12. Suppose $A, B, C \in 2^X$. Then

1. $d(A, B) = d(B, A)$
2. $d(A, A) = 0$
3. $d(A, B) + d(B, C) \geq d(A, C)$

Proof.

Lemma 13. **1.** $S(A, B) = S(B, A)$
  2. $S(A, A) = \emptyset$
  3. $S(A, B) \cup S(B,C) \supseteq S(A, C)$

Proof. Items 1 and 2 are obvious. To see item 3 we have

$$S(A, B) = (A - B) \cup (B - A)$$

and $$S(B, C) = (B - C) \cup (C - B)$$

so $$S(A, B) \cup S(B, C) = (A - B) \cup (B - A) \cup (B - C) \cup (C - B)$$

But $$A - C \subseteq (A - B) \cup (B - C)$$

and
$$C - A \subseteq (B - A) \cup (C - B)$$

so
$$S(A, C) = (A - C) \cup (C - A) \subseteq S(A, B) \cup S(B, C) \qquad \triangledown$$

The proposition follows from the lemma. $\qquad \square$

Note. $d(A, B) = 0$ if $\mu^*(S(A, B)) = 0$; that is, A and B symmetrically differ by a set of outer measure zero.

Although the preceding note says that d is not quite a distance function in the standard sense, we can still use d to define the notion of convergence in 2^X. That is, we say a sequence $\{A_i\}_{i=1}^{\infty} \in 2^X$ *converges* to $A \in 2^X$, written $A_i \to A$, if $d(A_i, A) \to 0$.

Proposition 14. The Boolean operations in 2^X are continuous with respect to d. That is, if $A_n \to A$ and $B_n \to B$ in 2^X, then

$$A_n \cup B_n \to A \cup B$$

$$A_n \cap B_n \to A \cap B$$

$$A_n - B_n \to A - B$$

and
$$A_n^c \to A^c$$

Proof.
Lemma 15. If $A_1, A_2, B_1, B_2 \in 2^X$, then

1. $S(A_1^c, B_1^c) = S(A_1, B_1)$
2. $S(A_1 \cup A_2, B_1 \cup B_2) \subseteq S(A_1, B_1) \cup S(A_2, B_2)$
3. $S(A_1 \cap A_2, B_1 \cap B_2) \subseteq S(A_1, B_1) \cup S(A_2, B_2)$
4. $S(A_1 - A_2, B_1 - B_2) \subseteq S(A_1, B_1) \cup S(A_2, B_2)$

Proof.

1. $S(A, B) = (A - B) \cup (B - A) = (A \cap B^c) \cup (B \cap A^c)$
   so $S(A^c, B^c) = (A^c \cap B) \cup (B^c \cap A) = S(A, B)$
2. $S(A_1 \cup A_2, B_1 \cup B_2) = [(A_1 \cup A_2) - (B_1 \cup B_2)] \cup [(B_1 \cup B_2) - (A_1 \cup A_2)]$
   $$= [(A_1 \cup A_2) \cap (B_1 \cup B_2)^c] \cup [(B_1 \cup B_2) \cap (A_1 \cup A_2)^c]$$
   $$= [(A_1 \cup A_2) \cap (B_1^c \cap B_2^c)] \cup [(B_1 \cup B_2) \cap (A_1^c \cap A_2^c)]$$
   $$\subseteq (A_1 \cap B_1^c) \cup (A_2 \cap B_2^c) \cup (B_1 \cap A_1^c) \cup (B_2 \cap A_2^c)$$
   $$= S(A_1, B_1) \cup S(A_2, B_2)$$
3. $S(A_1 \cap A_2, B_1 \cap B_2) = S(A_1^c \cup A_2^c, B_1^c \cup B_2^c)$
   $$\subseteq S(A_1^c, B_1^c) \cup S(A_2^c, B_2^c)$$
   $$= S(A_1, B_1) \cup S(A_2, B_2)$$
4. $S(A_1 - A_2, B_1 - B_2) = S(A_1 \cap A_2^c, B_1 \cap B_2^c)$
   $$\subseteq S(A_1, B_1) \cup S(A_2^c, B_2^c)$$
   $$= S(A_1, B_1) \cup S(A_2, B_2) \qquad \triangledown$$

From the lemma we have the immediate corollary

1. $d(A, B) = d(A^c, B^c)$
2. $d(A_1 \cup A_2, B_1 \cup B_2) \leq d(A_1, B_1) + d(A_2, B_2)$
3. $d(A_1 \cap A_2, B_1 \cap B_2) \leq d(A_1, B_1) + d(A_2, B_2)$
4. $d(A_1 - A_2, B_1 - B_2) \leq d(A_1, B_1) + d(A_2, B_2)$

from which the proposition follows.   □

Proposition 16. μ^* is continuous in the following sense: Let $A, B \in 2^X$ and suppose either $\mu^*(A)$ or $\mu^*(B)$ is finite, then

(19) $$|\mu^*(A) - \mu^*(B)| \leq d(A, B)$$

Proof.   Suppose $\mu^*(B) < \infty$; also assume $\mu^*(B) < \mu^*(A)$. Then

$$\mu^*(A) = d(A, \varnothing)$$

$$\leq d(B, \varnothing) + d(B, A)$$

$$= \mu^*(B) + d(B, A)$$

Thus

$$|\mu^*(A) - \mu^*(B)| = \mu^*(A) - \mu^*(B) \leq d(B, A)$$ □

Definition 17.   Let $\mathscr{M}_F$ be the closure of $\mathscr{R}$ in 2^X. That is, $A \in \mathscr{M}_F$ if and only if there exists a sequence of sets $\{A_i\}_{i=1}^\infty \subset \mathscr{R}$ such that $d(A_i, A) \to 0$ as $i \to \infty$.

Theorem 18.   1. $\mathscr{M}_F$ is a ring.
            2.   For $A \in \mathscr{M}_F$, $\mu^*(A) < \infty$.
            3. μ^* is a measure on $\mathscr{M}_F$.

Proof.

1.   Assume $A, B \in \mathscr{M}_F$. We need to show that $A \cup B$ and $A - B$ are in $\mathscr{M}_F$. Now, because $A, B \in \mathscr{M}_F$ there are sequences $\{A_i\}_{i=1}^\infty$ and $\{B_i\}_{i=1}^\infty$ in $\mathscr{R}$ such that $A_i \to A$ and $B_i \to B$. By the continuity of the Boolean operations

$$A_i \cup B_i \to A \cup B$$

$$A_i - B_i \to A - B$$

so $\mathscr{M}_F$ is a ring.
2. $A \in \mathscr{M}_F$ implies that there is a sequence $\{A_i\}_{i=1}^\infty \subset \mathscr{R}$ with $A_i \to A$. For some n, then, $d(A_n, A) < 1$.

Thus

$$\mu^*(A) \le \mu^*(A_n) + 1 < \infty$$

3. We first show that μ^* is additive, or—what amounts to the same thing—we will prove the lattice property; that is, if $A, B \in \mathcal{M}_F$ then

(20) $$\mu^*(A \cup B) + \mu^*(A \cap B) = \mu^*(A) + \mu^*(B)$$

Choose $A_n \to A$ and $B_n \to B$ in $\mathcal{R}$. Because on $\mathcal{R}$, $\mu^* = \mu$, μ^* is additive on $\mathcal{R}$. Thus

$$\mu^*(A_n \cup B_n) + \mu^*(A_n \cap B_n) = \mu^*(A_n) + \mu^*(B_n)$$

But $A_n \cup B_n \to A \cup B$ and $A_n \cap B_n \to A \cap B$, so the continuity of μ^* implies

$$\mu^*(A \cup B) + \mu^*(A \cap B) = \mu^*(A) + \mu^*(B)$$

We now prove countable additivity. Let $\{A_i\}_{i=1}^{\infty}$ be a mutually disjoint sequence in $\mathcal{M}_F$ with $A = \bigcup_{i=1}^{\infty} A_i$ also in $\mathcal{M}_F$. By the subadditivity of μ^* we know that

(21) $$\mu^*(A) \le \sum_{i=1}^{\infty} \mu^*(A_i)$$

Furthermore

$$\bigcup_{i=1}^{N} A_i \subset A$$

so $$\mu^*(A) \ge \mu^*\left(\bigcup_{i=1}^{N} A_i\right) = \sum_{i=1}^{N} \mu^*(A_i) \qquad \text{for all } N$$

That is,

(22) $$\mu^*(A) \ge \sum_{i=1}^{\infty} \mu^*(A_i) \qquad\qquad \square$$

Definition 19. A is a *measurable set*, $A \in \mathcal{M}$, if there exist $\{A_i\}_{i=1}^{\infty} \subset \mathcal{M}_F$ such that $A = \bigcup_{i=1}^{\infty} A_i$.

Theorem 20. If $A \in \mathcal{M}$ then $A \in \mathcal{M}_F \Leftrightarrow \mu^*(A) < \infty$.

Proof. Part 2 of theorem 18 gives "$\Rightarrow$", so to establish the theorem we must show that, if $\mu^*(A) < \infty$ and $A \in \mathcal{M}$, then $A \in \mathcal{M}_F$.

Because $A \in \mathcal{M}$, there exist $A_i \in \mathcal{M}_F$ such that $A = \bigcup_{i=1}^{\infty} A_i$. We can assume this union is disjoint for, if it isn't, we can replace the A_i's by $\tilde{A}_i$'s as follows:

$$\tilde{A}_1 = A_1$$

$$\tilde{A}_2 = A_2 - A_1$$

$$\tilde{A}_3 = A_3 - (A_1 \cup A_2)$$

and, because $\mathcal{M}_F$ is a ring, we know $\tilde{A}_i \in \mathcal{M}_F$. Thus we can assume $A = \bigcup_{i=1}^{\infty} A_i$ is a disjoint union.

Now consider $\mu^*(A)$. First, subadditivity gives $\mu^*(A) \le \sum_{i=1}^{\infty} \mu^*(A_i)$. We claim that, in fact, $\mu^*(A) = \sum_{i=1}^{\infty} \mu^*(A_i)$. To see this, notice that

$$\bigcup_{i=1}^{N} A_i \subset A$$

so
$$\mu^*\left(\bigcup_{i=1}^{N} A_i\right) = \sum_{i=1}^{N} \mu^*(A_i) \le \mu^*(A)$$

Because this equality holds for any N, we have

$$\sum_{i=1}^{\infty} \mu^*(A_i) \le \mu^*(A)$$

and thus

(23)
$$\sum_{i=1}^{\infty} \mu^*(A_i) = \mu^*(A)$$

Now, fix $\varepsilon > 0$ and let $B_N = \bigcup_{i=1}^{N} A_i$; then $B_N \in \mathcal{M}_F$, and

(24)
$$d(A, B_N) = \mu^*(A - B_N) = \mu^*\left(\bigcup_{j>N} A_j\right)$$

$$\le \sum_{j>N} \mu^*(A_j) < \varepsilon \qquad \text{for } N \text{ large}$$

because $\sum_{i=1}^{\infty} \mu^*(A_i)$ is convergent. Thus $A \in \mathcal{M}_F$ because $B_N \to A$ and $\mathcal{M}_F$ is closed. $\qquad\square$

We now consider properties of the collection $\mathcal{M}$.

Definition 21. Let $\mathcal{S}$ be a collection of subsets of a set X. $\mathcal{S}$ is called a σ-ring if

1. it is a ring and
2. given $\{A_i\}_{i=1}^{\infty}$ in $\mathcal{S}$, $\bigcup_{i=1}^{\infty} A_i$ is also in $\mathcal{S}$.

Theorem 22. $\mathcal{M}$ is a σ-ring.

Proof. First we will show property 2.

Suppose $A_1, A_2, \ldots$ are elements of $\mathcal{M}$. Let $A = \bigcup_{i=1}^{\infty} A_i$. Because each $A_i \in \mathcal{M}$, there must be $\{A_{ij}\}_{j=1}^{\infty}$ in $\mathcal{M}_F$ such that

$$A_i = \bigcup_{j=1}^{\infty} A_{ij}$$

Then $A = \bigcup_{i,j=1}^{\infty} A_{ij}$, a countable union, so $A \in \mathscr{M}$.

Now we will show that $\mathscr{M}$ is a ring. It suffices to show that if $A, B \in \mathscr{M}$ then $A - B \in \mathscr{M}$.

First, suppose $A \in \mathscr{M}_F$ and write $B = \bigcup_{i=1}^{\infty} B_i$ with $B_i \in \mathscr{M}_F$. Because $\mathscr{M}_F$ is a ring, $A \cap B_i \in \mathscr{M}_F$ so $A \cap B = \bigcup_{i=1}^{\infty} A \cap B_i$ is a member of $\mathscr{M}$. Moreover, $\mu^*(A \cap B) \leq \mu^*(A) < \infty$, so $A \cap B \in \mathscr{M}_F$. Now $A - B = A - (A \cap B)$ and, because $\mathscr{M}_F$ is a ring of which A and $A \cap B$ are members, we have $A - B \in \mathscr{M}_F$.

Now let A be a general element of $\mathscr{M}$ and write $A = \bigcup_{i=1}^{\infty} A_i$ with $A_i \in \mathscr{M}_F$. Then

$$A - B = \bigcup_{i=1}^{\infty} (A_i - B)$$

but from the discussion above, $A_i - B \in \mathscr{M}_F$, so we are done. $\qquad \square$

Theorem 23. If $A_1, A_2, \ldots$ is a countable collection of disjoint sets in $\mathscr{M}$, then

$$\mu^*\left(\bigcup_{i=1}^{\infty} A_i\right) = \sum_{i=1}^{\infty} \mu^*(A_i)$$

Proof. Let $A = \bigcup_{i=1}^{\infty} A_i$, $A \in \mathscr{M}$. We consider two cases separately.

1. $\mu^*(A) < \infty$

   Because $A_i \subset A$, $\mu^*(A_i) < \infty$ so A and all of the A_i's are elements of $\mathscr{M}_F$. Because μ^* is a measure on $\mathscr{M}_F$, we have then

   $$\mu^*(A) = \sum_{i=1}^{\infty} \mu^*(A_i)$$

2. $\mu^*(A) = \infty$

   In this case subadditivity tells us that

   $$\infty = \mu^*(A) \leq \sum_{i=1}^{\infty} \mu^*(A_i)$$

   so

   $$\sum_{i=1}^{\infty} \mu^*(A_i) = \infty \qquad \square$$

Now that we have constructed measurable sets in the abstract case, let us return to the example of Lebesgue measure in $\mathbf{R}^n$.

Example 24. $X = \mathbf{R}^n$, $\mathscr{R} = \mathscr{R}_{\text{Leb}}$ = finite unions of multi-intervals, and μ is as given in example 4. Here we call $\mathscr{M}$ the set of Lebesgue measurable sets in

$\mathbf{R}^n$, and the extension of μ to $\mathcal{M}$ (the restriction of μ^* to $\mathcal{M}$) is called Lebesgue measure μ_L.

What do the sets in $\mathcal{M}$ look like? First we remark that $\mathbf{R}^n \in \mathcal{M}$. Indeed let

$$I_N = \{x \in \mathbf{R}^n; \ -N \leq x_i \leq N, i = 1, \ldots, n\}$$

Then

$$\mathbf{R}^n = \bigcup_{N=1}^{\infty} I_N \qquad \text{and each } I_N \in \mathcal{R} \subset \mathcal{M}_F$$

Proposition 25.   Every open subset of $\mathbf{R}^n$ is in $\mathcal{M}$.

Proof.   Let $c = (a_1, a_2, \ldots, a_n, b_1, b_2, \ldots, b_n) \in \mathbf{R}^{2n}$ with the a_i's and b_i's rational and $a_i < b_i$. Let $I_c = \{x \in \mathbf{R}^n; \ a_i < x_i < b_i, i = 1, 2, \ldots, n\}$. The collection $\{I_c\}$ is countable.

Now let $\mathcal{O}$ be any open subset of $\mathbf{R}^n$; $\mathcal{O}$ is equal to the union of all sets I_c such that $I_c \subset \mathcal{O}$. (If $x \in \mathcal{O}$ we can find a c such that $x \in I_c \subset \mathcal{O}$.) Because such a union is countable, $\mathcal{O} \in \mathcal{M}$.

Corollary 26.   Every closed subset of $\mathbf{R}^n$ is in $\mathcal{M}$.

Proof. A is closed so A^c is open. $A = \mathbf{R}^n - A^c$ and, because $\mathcal{M}$ is a ring and $\mathbf{R}^n, A^c \in \mathcal{M}$, we have $A \in \mathcal{M}$.

Corollary 27.   All countable unions and intersections of closed and open sets are measurable.

We have shown that the measurable sets are a σ-ring containing the open subsets of $\mathbf{R}^n$. Are they the smallest σ-ring with this property? That is, if one starts with the closed and open sets, forms countable unions and intersections, and then from these forms further countable unions and intersections, and so on, does one eventually end up with *all* measurable sets? The answer is no, unfortunately; so we are forced to make the following definition.

Definition 28.   The *Borel sets* are the smallest σ-ring containing the open sets.

Although not all measurable sets are Borel (for an example see Halmos, P. *Measure Theory.* [Van Nostrand: Princeton, NJ] p. 67), the following theorem says that any measurable set is close to being a Borel set.

Theorem 29.   If $A \in \mathcal{M}$ there exists a Borel set $B \subseteq A$ such that $\mu^*(A - B) = 0$; that is, A can be written as $A = (A - B) \cup B$, where B is Borel and $\mu^*(A - B) = 0$.

Lemma 30. If $A \in \mathcal{M}$ and if $\varepsilon > 0$ is given, then there exists a Borel set G such that $G \supset A$ and $\mu^*(G - A) < \varepsilon$.

Proof. First suppose $\mu^*(A) < \infty$. Then by definition of μ^* we can find a cover $A \subset \bigcup_{i=1}^{\infty} A_i$ such that

$$\sum \mu(A_i) \leq \mu^*(A) + \varepsilon$$

where each of the A_i's is a multi-interval. But $G = \bigcup_{i=1}^{\infty} A_i$ is Borel.

More generally, if $A \in \mathcal{M}$ we can write $A = \bigcup_{i=1}^{\infty} A_i$ where each $A_i \in \mathcal{M}_F$. By the preceding argument we can find Borel sets G_i with $A_i \subset G_i$ and $\mu^*(G_i - A_i) < \varepsilon/2^i$. Then with $G = \bigcup_{i=1}^{\infty} G_i$ we have

$$\mu^*(G - A) < \varepsilon \qquad\qquad \triangledown$$

Lemma 31. If $A \in \mathcal{M}$ there exists a Borel set $F \subset A$ with

$$\mu^*(A - F) < \varepsilon$$

Proof. Choose a Borel set G such that $A^c \subset G$ and $\mu^*(G - A^c) < \varepsilon$ by lemma 30. Let $F = G^c$. Then $A - F = G - A^c$ and

$$\mu^*(A - F) = \mu^*(G - A^c) < \varepsilon \qquad\qquad \triangledown$$

Now we prove theorem 29. Take $A \in \mathcal{M}$. For every N choose a Borel set $F_N \subset A$ such that $\mu^*(A - F_N) < 1/N$. Let $F = \bigcup_{N=1}^{\infty} F_N$; then F is Borel and

$$\mu^*(A - F) \leq \mu^*(A - F_N) < \frac{1}{N} \qquad \text{for every } N$$

Thus $\mu^*(A - F) = 0$.   □

We conclude this section with a few remarks about notation. Let X be a set, $\mathcal{R}$ a ring of subsets of X, and μ a measure on $\mathcal{R}$. By theorem 18 μ extends to a measure on a much larger ring of sets, $\mathcal{M}_F$. In fact μ can be regarded as a measure on the σ-ring $\mathcal{M}$, providing we define it to take the value $+\infty$ on sets A that are in $\mathcal{M}$ but not in $\mathcal{M}_F$. Note that proposition 5 is then still true if one observes the usual addition conventions for $+\infty$, namely

$$(+\infty) + a = +\infty \qquad \text{for } a \in \mathbf{R}$$

and $$(+\infty) + (+\infty) = +\infty$$

Moreover, μ is countably additive on $\mathcal{M}$ by theorem 23. Note that, if X itself is in $\mathcal{M}_F$, these problems with infinity don't arise; that is, $\mathcal{M}_F = \mathcal{M}$. For all examples of measures that we will encounter in this text, the set X is either in $\mathcal{M}_F$ or in $\mathcal{M}$—that is, X satisfies the conditions of the following definition.

Definition 32. X is σ-finite if there exist sets $X_i \in \mathcal{M}_F$, $i = 1, 2, \ldots$, with $X = \bigcup_{i=1}^{\infty} X_i$.

For example, $\mathbf{R}^n$ is σ-finite because

$$\mathbf{R}^n = \bigcup_{i=1}^{\infty} B_i$$

with B_i being the ball of radius i about the origin.

Exercises for §1.3

1. Let X be an uncountable set. Let $\mathcal{R}$ be the collection of all finite subsets of X. Given $A \in \mathcal{R}$ let $\mu(A)$ be the number of elements in A. Show that $\mathcal{R}$ is a ring and that μ is a measure on $\mathcal{R}$. Identify μ^*. What are $\mathcal{M}$ and $\mathcal{M}_F$? Is every subset of X measurable?

2. Let X be an infinite set and let $\mathcal{R}$ be the following collection of subsets: $A \in \mathcal{R}$ if and only if A is finite or A^c is finite. Let μ be the following function on $\mathcal{R}$: $\mu(A) = 0$ if A is finite, and $\mu(A) = 1$ if A^c is finite. Is μ a measure?

3. **a.** Let X be an infinite set and $\mathcal{R}$ the collection of all countable subsets of X. Is $\mathcal{R}$ a σ-ring?

   b. Let μ be a measure on $\mathcal{R}$. Show that there exists a function $f: X \to [0, \infty)$ such that

   (∗) $$\mu(A) = \sum_{x \in A} f(x)$$

   for all $A \in \mathcal{R}$.

   c. Show that the function f in part b has to have the following two properties: (1) The set $\{x \in X; f(x) \neq 0\}$ is countable and (2) $\sum_{x \in X} f(x) < \infty$.

   d. Show that, if f has the properties in part c, the formula (∗) defines a measure on $\mathcal{R}$.

4. Let X be the real line and $\mathcal{R} = \mathcal{R}_{\text{Leb}}$. (That is, finite unions of intervals.) Given $A \in \mathcal{R}$ let $\mu(A) = 1$ if, for some positive ε, A contains the interval $(0, \varepsilon)$. Otherwise let $\mu(A) = 0$. Show that μ is an additive set function but is *not* countably additive.

5. Let F be a continuous, monotone increasing function on the real line. If A is an interval with endpoints a and b, let

   $$\mu_F(A) = F(b) - F(a)$$

   More generally, if A is a disjoint union of intervals

   $$A = \bigcup_{i=1}^{N} A_i$$

let $\mu_F(A) = \sum_{i=1}^{N} \mu_F(A_i)$. Show that μ_F is a measure on the ring $\mathscr{R}_{\text{Leb}}$; that is, prove it is countably additive.

Remark. If one takes for F an antiderivative of the function $(1/\sqrt{2\pi})e^{-x^2/2}$, μ_F is called the *Gaussian* measure. We will encounter it several times later on.

6. **a.** Let A be a measurable subset of **R**. One says that the *density of A is well defined* if the limit

$$D(A) = \lim_{T \to \infty} \frac{\mu_L\{A \cap [-T, T]\}}{2T}$$

exists. If the limit exists, this expression is called the density of A. Can you produce an example of a measurable set A whose density is *not* defined?

b. Show that, if A_1 and A_2 have well-defined densities and are disjoint, then $A_1 \cup A_2$ has a well-defined density and

$$D(A_1 \cup A_2) = D(A_1) + D(A_2)$$

c. Show that there exist sets A and A_i, $i = 1, 2, \ldots$, with well-defined densities such that

$$A = \bigcup_{i=1}^{\infty} A_i \qquad \text{(disjoint unions)}$$

but

$$D(A) \neq \sum D(A_i)$$

7. Let X be a set, $\mathscr{R}$ a σ-ring of subsets of X, and μ_1 and μ_2 measures on $\mathscr{R}$. Let $\mathscr{L}$ be the family of all those sets $A \in \mathscr{R}$ for which $\mu_1(A) = \mu_2(A)$. Assume $X \in \mathscr{R}$ and $\mu_1(X) = \mu_2(X) < \infty$. Show that $\mathscr{L}$ has the following properties:

(**)

    (i) $X \in \mathscr{L}$.
    (ii) If $A, B \in \mathscr{L}$ and $B \subseteq A$, $A - B \in \mathscr{L}$.
    (iii) If $A_i \in \mathscr{L}$, $i = 1, 2, \ldots$ and $A = \bigcup_{i=1}^{\infty} A_i$ (disjoint union)
    then $A \in \mathscr{L}$.

Remark. A collection of sets $\mathscr{L}$ having the properties listed in (**) is called a λ-system.

8. Let X be the three-element set $\{P_1, P_2, P_3\}$, and let $\mathscr{R}$ be the ring of subsets of X. Let μ_1 and μ_2 be measures on $\mathscr{R}$. When is the set $\mathscr{L}$ a ring? Show that $\mathscr{L}$ doesn't always have to be a ring.

9. Show that the example described in exercise 1 is not σ-finite.

10. Remember that a metric space is *complete* if every Cauchy sequence

has a limit. Show that, with respect to the distance function $d(A, B) = \mu^*(S(A, B))$, 2^X is complete.

11. **a.** Given any collection $\mathscr{C}$ of subsets of a set X, show that there is a *smallest* ring of sets $\mathscr{R}$ containing $\mathscr{C}$. (That is, $\mathscr{R}$ has the property that it contains $\mathscr{C}$, and *any* ring that contains $\mathscr{C}$ contains $\mathscr{R}$.) Describe explicitly how to construct $\mathscr{R}$ from $\mathscr{C}$.

    b. Show that the ring $\mathscr{R}_{\text{Leb}}$ is the smallest ring containing the multi-intervals.

12. Given any collection $\mathscr{C}$ of subsets of a set X, show that there exists a *smallest* σ-ring of sets $\mathscr{R}_\sigma$ containing $\mathscr{C}$. This justifies definition 28.

13. Show that for any $\delta > 0$ there exists an open dense subset U of $\mathbf{R}$ with $\mu_L(U) < \delta$.

14. Let $c \in \mathbf{R}^n$. Given any subset A of $\mathbf{R}^n$, let $A + c = \{w \in \mathbf{R}^n; w - c \in A\}$. Prove that, if A is measurable, then $A + c$ is measurable and

$$(**) \qquad \mu_L(A + c) = \mu_L(A)$$

(*Hint:* First, prove this for multi-intervals. Next, show that in equation $(**)$ the outer measures are equal.)

15. Let $f: \mathbf{R} \to \mathbf{R}$ be the linear mapping $x \to ax + b$, a and b being constants with $a > 0$. Show that, if A is measurable, $f(A)$ is measurable and

$$\mu_L(f(A)) = a\mu_L(A)$$

16. Let A be a Lebesgue measurable subset of $\mathbf{R}$ and let

$$C_A = \{(x, y) \in \mathbf{R}^2; x \in A\}$$

Such sets are called cylinder sets. Show that the collection of these sets forms a ring $\mathscr{R}_C$. Show that the set function μ_C defined by

$$\mu_C(C_A) = \mu_L(A)$$

is a measure on this ring. Show that, if S is a proper subset of $\mathbf{R}$ and $\mu_L(A) \neq 0$, the set

$$A \times S = \{(x, y); x \in A, y \in S\}$$

is not a measurable subset of $\mathbf{R}^2$ with respect to the measure μ_C. (*Hint:* What is its outer measure, computed with respect to μ_C?)

17. Let $f: \mathbf{R}^m \to \mathbf{R}^n$ be a continuous map. Show that, if A is a Borel subset of $\mathbf{R}^n$, then $f^{-1}(A)$ is a Borel subset of $\mathbf{R}^m$. Define

$$\mu_f(A) = \mu_L(f^{-1}(A))$$

Show that μ_f is a measure on the Borel subsets of $\mathbf{R}^n$.

18. Let $\mathscr{R}$ be a ring and μ a measure on $\mathscr{R}$. Prove that, if $A_1, A_2, \dots, A_n$ are in $\mathscr{R}$ then

$$\mu(A_1 \cup \cdots \cup A_n) = \sum_{i=1}^{n} \mu(A_i) - \sum_{i<j} \mu(A_i \cap A_j)$$

$$+ \sum_{i<j<k} \mu(A_i \cap A_j \cap A_k) + \cdots$$

$$+ (-1)^{n+1} \mu(A_1 \cap \cdots \cap A_n)$$

19. Let X be a set, $\mathscr{R}$ a ring of subsets of X, and μ a measure on $\mathscr{R}$. Let μ^* be the corresponding outer measure. Show that if $A \in \mathscr{M}_F$ then, for all $E \subseteq X$,

(†) $$\mu^*(A \cap E) + \mu^*(A^c \cap E) = \mu^*(E)$$

(*Hint:* First check this for $A \in \mathscr{R}$.)

20. Prove the converse result; that is, suppose $\mu^*(A)$ is finite and A satisfies property (†) for every subset E of X. Prove that $A \in \mathscr{M}_F$. (*Hint:* Show that, for every set A and every $\varepsilon > 0$, there exists a set $E \in \mathscr{M}$ such that $E \supset A$ and $\mu^*(E) \leq \mu^*(A) + \varepsilon$. Use property (†) to conclude that $d(E, A) < \varepsilon$.)

Remark. In many textbooks the property (†) is used as the *definition* of a measurable set.

§1.4 Measure Theoretic Modeling

Now that we have developed the basic notions of measure theory we can examine a little more closely the ideas involved in what we have called the "Borel principle." First we provide some definitions.

Definition 1. Let X be a set and $\mathscr{F}$ a ring of subsets of X.

1. $\mathscr{F}$ is a *field* if $X \in \mathscr{F}$.
2. $\mathscr{F}$ is a *σ-field* if $X \in \mathscr{F}$ and if $\mathscr{F}$ is a *σ-ring*.

Definition 2. Let X be a set and $\mathscr{F}$ a field of subsets of X. Suppose μ is a measure defined on $\mathscr{F}$. Then μ is a *probability measure* if $\mu(X) = 1$. In this case the triple $\{X, \mathscr{F}, \mu\}$ is called a *probability space*.

Example 3. Let X be the unit interval I, and let $\mathscr{F}$ be the measurable subsets contained in I. Then $\mathscr{F}$ is a σ-field and the Lebesgue measure is a probability measure.

Now let X be the sample space of a probabilistic process. A *measure theoretic model* of the process is a σ-field $\mathscr{F}$ of subsets of X and a probability measure μ defined on $\mathscr{F}$ so that, for any "plausible" event E in X, we have

$B_E \in \mathcal{F}$ and $\text{Prob}(E) = \mu(B_E)$, where B_E is the set of points in X for which E occurs.

Of course this definition is not precise; the word *plausible* is left to be interpreted by the modeler. We need to include the plausibility qualification because the desired measure may not be defined on all subsets of X. For example, as we remarked parenthetically in §1.3, not every subset of the unit interval is Lebesgue measurable.

Let us consider some examples of measure theoretic models.

I.   Discrete Probability Theory

Suppose the sample space X is finite or countable, say $X = \{x_1, x_2, x_3, \ldots\}$. Further, suppose that each point x_i has the probability p_i of occurring and that $\sum p_i = 1$. The measure theoretic model for this process is given by letting $\mathcal{F}$ be the collection of all subsets of X and by defining μ as

$$(1) \qquad\qquad \mu(A) = \sum_{x_i \in A} p_i \qquad \text{for } A \subset X$$

It is left as an exercise for the reader to check that μ is a measure. (See exercise 3 in §1.3.)

Notice that in this case we need not interpret the word *plausible* because $\mathcal{F}$ contains all subsets of X; that is, all events are considered plausible.

II.   Bernoulli Sequences and Random Walks

In this case the sample space can be identified with the unit interval I, and the measure theoretic model is given by the Borel principle. In §1.2 we saw that, for many "plausible" events E, B_E is a finite union of intervals (and thus measurable) and that $\text{Prob}(E) = \mu(B_E)$ in these cases. The events considered there were rather simple; let us now confirm that $B_E \in \mathcal{F}$ for some more complicated, yet still "plausible," events.

1. Let E be the event that a prescribed finite pattern, for example, H T T H, occurs infinitely often. To describe B_E we let E_n be the event that the pattern occurs beginning at the nth step. Because B_{E_n} is described by a finite number of conditions on the Rademacher functions, it is a finite union of intervals. (If the pattern is H T T H, then $B_{E_n} = \{\omega \in I; \; R_n(\omega) = 1, \; R_{n+1}(\omega) = -1, \; R_{n+2}(\omega) = -1, R_{n+3}(\omega) = 1\}$.) Thus, because

$$(2) \qquad\qquad B_E = \bigcap_{k=1}^{\infty} \bigcup_{n \geq k} B_{E_n}$$

it is Borel.

2. (Law of large numbers) Let E be the event that a Bernoulli sequence obeys the law of large numbers. That is, with $S_n(\omega) = \sum_{k=1}^{n} R_k(\omega)$,

$$B_E = \left\{\omega \in I; \frac{S_n(\omega)}{n} \to 0 \text{ as } n \to \infty\right\}$$

We know this is measurable because we showed its complement has measure zero. However, let us describe B_E as a Borel set.

Recall that the statement

$$\frac{S_n(\omega)}{n} \to 0 \quad \text{as} \quad n \to \infty$$

means that, for every integer $r > 0$, there is a $k > 0$ such that

$$\left|\frac{S_n(\omega)}{n}\right| < \frac{1}{r} \qquad \text{whenever } n \geq k$$

If we let

(3)
$$A_{n,r} = \left\{\omega \in I; \left|\frac{S_n(\omega)}{n}\right| < \frac{1}{r}\right\}$$

we can write

(4)
$$B_E = \bigcap_{r=1}^{\infty} \bigcup_{k=1}^{\infty} \bigcap_{n \geq k} A_{n,r}$$

which is Borel, because each $A_{n,r}$ is a finite union of intervals.

3. Let E be the event that $\sum (R_n(\omega)/n)$ converges. We claim B_E is Borel. Let

(5)
$$T_n(\omega) = \sum_{k=1}^{n} \frac{R_n(\omega)}{n}$$

Then the Cauchy condition tells us that $\sum (R_n(\omega)/n)$ converges if, for every integer $r > 0$, there is a $k > 0$ such that

$$|T_n(\omega) - T_m(\omega)| < \frac{1}{r} \qquad \text{for all } m, n \geq k$$

If we let

$$A_{m,n,r} = \left\{\omega \in I; |T_m(\omega) - T_n(\omega)| < \frac{1}{r}\right\}$$

we see that

(6)
$$B_E = \bigcap_{r=1}^{\infty} \bigcup_{k=1}^{\infty} \bigcap_{m,n \geq k} A_{m,n,r}$$

which is Borel, because $A_{m,n,r}$ is a finite union of intervals.

We have now shown that these "plausible" events correspond to measurable sets. Thus, if we assume that the Borel principle holds, we can determine the probability of these events by finding the measure of these sets. We know already that, for the law of large numbers, the set described has measure one. We now develop the necessary tools to determine the measure of the set described in event **1**.

This example is a special case of the following general situation: Start with a countable collection of events

$$\{E_1, E_2, \ldots\}$$

and define a new event E to be the event that infinitely many of the events E_i occur. Can we determine $\text{Prob}(E)$ if we know $\text{Prob}(E_i)$ for all i? Two theorems address this problem; they are called the Borel–Cantelli lemmas. In order to formulate them, we first restate the problem in measure theoretic terms.

Let X be the sample space of our process, equipped with a σ-field $\mathscr{F}$ and a measure μ. Let B_i denote the subset of X on which E_i occurs. We assume that $B_i \in \mathscr{F}$ and that $\text{Prob}(E_i) = \mu(B_i)$. If we let B_E denote the subset of X corresponding to the event E, then, in terms of the B_i's,

$$(7) \qquad B_E = \bigcap_{k=1}^{\infty} \bigcup_{n \geq k} B_n$$

We give this a name.

Definition 4. Given sets $B_1, B_2, B_3, \ldots$ in $\mathscr{F}$, then

$$(8) \qquad \{B_i; \text{i.o.}\} = \limsup B_n = \bigcap_{k=1}^{\infty} \bigcup_{n \geq k} B_n$$

is called "B_i, infinitely often" or the *limes supremum* of the B_i's.

Theorem 5. (First Borel–Cantelli lemma) Given $B_1, B_2, \ldots$ in $\mathscr{F}$, let $B = \{B_i; \text{i.o.}\}$. Then $\sum_{i=1}^{\infty} \mu(B_i) < \infty$ implies that $\mu(B) = 0$.

Proof. Let $A_k = \bigcup_{n \geq k} B_n$ so that $B = \bigcap_{k=1}^{\infty} A_k$; in particular, $B \subseteq A_k$ for all k. Now, by subadditivity

$$\mu(A_k) \leq \sum_{n \geq k} \mu(B_n)$$

Thus, because $\sum_{n=1}^{\infty} \mu(B_n) < \infty$, for $\varepsilon > 0$ there is a $k > 0$ such that

$$\mu(A_k) \leq \sum_{n \geq k} \mu(B_n) < \varepsilon$$

Because $B \subseteq A_k$ we have that $\mu(B) < \varepsilon$, and because ε is arbitrary we must have $\mu(B) = 0$. $\qquad\square$

Application. (Run lengths) For $\omega \in I$ define the nth run-length function l_n by letting $l_n(\omega)$ be the number of consecutive 1's in the binary expansion of ω starting at the nth place. That is, $l_n(\omega) = k$ if $R_n(\omega) = 1$, $R_{n+1}(\omega) = 1, \ldots, R_{n+k-1}(\omega) = 1$, and $R_{n+k}(\omega) = -1$.

Now take a sequence of numbers, $r_1, r_2, r_3, \ldots$, and let E_n denote the event that $l_n(\omega) \geq r_n$. Let $E = \{E_n; \text{i.o.}\}$. Then

$$B_{E_n} = \{\omega \in I; R_n(\omega) = R_{n+1}(\omega) = \cdots = R_{n+r_n-1}(\omega) = 1\}$$

so $\mu(B_{E_n}) = (\frac{1}{2})^{r_n}$ and we can use theorem 5 to conclude the following.

Corollary. If $\sum_{n=1}^{\infty} (1/2)^{r_n} < \infty$ then $\mu(B_E) = 0$.

The second Borel–Cantelli lemma supplies a partial converse to the first. It is restricted by applying only to independent events.

Definition 6. Two events E_1 and E_2 are *independent* if the outcome of E_1 tells us nothing about the outcome of E_2.

Let us try to make this definition more precise by restating it in measure theoretic terms. Knowing that the event E_1 occurs means that the elements of the sample space in which we are interested are already in B_{E_1}. Now, for what proportion of the elements in B_{E_1} does the event E_2 occur? Clearly the answer is

(9)
$$\frac{\mu(B_{E_1} \cap B_{E_2})}{\mu(B_{E_1})}$$

This ratio is called the *conditional probability* of E_2 given E_1. Now, if E_2 is independent of E_1, this conditional probability is just the probability of E_2 computed without prior knowledge of E_1—that is, $\mu(B_{E_2})$; hence

$$\mu(B_{E_2}) = \frac{\mu(B_{E_1} \cap B_{E_2})}{\mu(B_{E_1})}$$

This leads us to the following measure theoretic definition.

Definition 7. Let X be a sample space with σ-field $\mathscr{F}$ and probability measure μ. Two sets $A_1, A_2 \in \mathscr{F}$ are *independent* if

(10)
$$\mu(A_1 \cap A_2) = \mu(A_1)\mu(A_2)$$

Example 8. Given $X = I$, $\mu =$ Lebesgue measure, $A_1 = \{\omega \in I; R_1(\omega) = 1\}$, and $A_2 = \{\omega \in I; R_2(\omega) = 1\}$, then

$$A_1 = (\tfrac{1}{2}, 1] \quad \text{and} \quad A_2 = (\tfrac{1}{4}, \tfrac{1}{2}] \cup (\tfrac{3}{4}, 1]$$

$$A_1 \cap A_2 = (\tfrac{3}{4}, 1]$$

Thus, $\qquad\qquad \mu(A_1 \cap A_2) = \tfrac{1}{4} = (\tfrac{1}{2})^2 = \mu(A_1)\mu(A_2)$

Definition 9.   More generally, $A_1, A_2, \ldots, A_n$ are *independent* if, for any sequence of integers $1 \le i_1 < i_2 < \cdots < i_k \le n$, we have

(11) $\qquad\qquad \mu(A_{i_1} \cap A_{i_2} \cap \cdots \cap A_{i_k}) = \mu(A_{i_1})\mu(A_{i_2})\cdots\mu(A_{i_k})$

Further, a countable collection of sets is *independent* if every finite subcollection is independent.

Example 10.   Let $A_i = \{\omega \in I; R_i(\omega) = 1\}$. It is left as an exercise to prove that the collection $A_1, A_2, \ldots$ is independent.

Theorem 11.   (Second Borel–Cantelli lemma) Let $(X, \mathcal{F}, \mu)$ be a probability space and let $A_1, A_2, \ldots$ be an independent collection of sets from $\mathcal{F}$. Suppose that $\sum \mu(A_i) = \infty$; then $\mu(\{A_i; \text{i.o.}\}) = 1$.

Lemma 12.   Let $A_1, A_2, \ldots$ be an independent collection of sets in $\mathcal{F}$. Then $A_1^c, A_2^c, A_3^c, \ldots$ is an independent collection of sets in $\mathcal{F}$.
   The proof of the lemma is left to the reader. (See exercise 10. We suggest that the reader give this exercise a few moments of thought before continuing the chapter.)

Proof of theorem.   Let $A = \{A_i; \text{i.o.}\}$. Then

$$A = \bigcap_{k=1}^{\infty} \bigcup_{n \ge k} A_n \qquad \text{so} \qquad A^c = \bigcup_{k=1}^{\infty} \bigcap_{n \ge k} A_n^c$$

To show that $\mu(A) = 1$, it is enough to show that $\mu(A^c) = 0$; and, to establish this fact it is enough to show (by subadditivity) that

$$\mu\left(\bigcap_{n \ge k} A_n^c\right) = 0$$

Now, by independence,

$$\mu\left(\bigcap_{n=k}^{l} A_n^c\right) = \prod_{n=k}^{l} \mu(A_n^c)$$

but $\mu(A_n^c) = 1 - \mu(A_n)$, which in turn is less than or equal to $e^{-\mu(A_n)}$ because it is true in general that $1 - x \le e^{-x}$. (Prove this inequality yourself!) Thus

(12)
$$\mu\left(\bigcap_{n=k}^{l} A_n^c\right) \le \prod_{n=k}^{l} e^{-\mu(A_n)} = e^{-\sum_{n=k}^{l} \mu(A_n)}$$

But $e^{-\sum_{n=k}^{l} \mu(A_n)} \to 0$ as $l \to \infty$, because $\sum_{n=1}^{\infty} \mu(A_n) = \infty$. Thus $\mu(\bigcap_{n \ge k} A_n^c) = 0$. $\square$

Example 13.   Let H_n denote the event of a head at the nth toss of a Bernoulli sequence. The corresponding subset of I is

$$A_n = \{\omega \in I; \; R_n(\omega) = 1\}$$

In exercise 11 the reader will show that these sets are independent. Furthermore, $\mu(A_n) = \frac{1}{2}$ for each n, so $\sum \mu(A_n) = \infty$. Thus a head occurs infinitely often in a Bernoulli sequence with probability one. (This result can be proved much more trivially. What is the proof?)

Example 14.   Example 13 is an example of a finite pattern (the pattern H) occurring infinitely often in Bernoulli sequences. More generally we now show that any finite pattern occurs infinitely often in Bernoulli sequences with probability one. For simplicity of notation, consider the particular pattern H T T H.

Proposition 15.   The pattern H T T H occurs infinitely often in a Bernoulli sequence with probability one.

Proof.   Let E_n be the event that H T T H occurs starting at step n, and let B_n be the corresponding subset of I. Because the A_n's are independent and $B_n = A_n \cap A_{n+1}^c \cap A_{n+2}^c \cap A_{n+3}$, we have $\mu(B_n) = (\frac{1}{2})^4 = \frac{1}{16}$; so $\sum_{n=1}^{\infty} \mu(B_n) = \infty$. Unfortunately, B_n and B_{n+1} are not independent, so the second Borel–Cantelli lemma does not apply. However, the sets $B_n, B_{n+4}, B_{n+8}, \ldots$ are independent; in particular, $B_1, B_5, B_9, \ldots, B_{4k+1}, \ldots$ are independent and

$$\sum_{k=1}^{\infty} \mu(B_{4k+1}) = \infty$$

so the second Borel–Cantelli lemma applies to give

$$\mu(\{B_{4k+1}; \text{i.o.}\}) = 1$$

But $\{B_{4k+1}; \text{i.o.}\} \subset \{B_n; \text{i.o.}\}$ so

$$1 = \mu(\{B_{4k+1}; \text{i.o.}\}) \le \mu(\{B_n; \text{i.o.}\}) \le 1$$

Thus $\mu(\{B_n; \text{i.o.}\}) = 1$. $\square$

Remark. This same proof works for any finite pattern of H's and T's—for example, Shakespeare's sonnets translated into Morse code, with the dots and dashes changed to H's and T's.

Exercises for §1.4

1. We will say that an event E involving Bernoulli sequences is plausible if the subset B_E of the unit interval corresponding to it is a Borel subset. Show that the following events are plausible:
   a. A gambler quadruples his initial stake. (*Beware:* he will not quadruple his initial stake if he gets wiped out beforehand.)
   b. In an infinite sequence of trials, a gambler breaks even an infinite number of times.
   c. In an infinite sequence of trials, arbitrarily long run lengths occur.
   d. In an infinite sequence of trials, H comes up "on the average" more often than T.
   (Incidentally, event **d** shows that "plausible" does not necessarily mean "probable.")

2. Show that, for random walks on the line, the following events are plausible:
   a. The origin is visited infinitely often.
   b. Every integer point on the real line is visited infinitely often.

3. Show that, for random walks in the plane, the following events are plausible:
   a. The origin is visited infinitely often.
   b. Every point (m, n) is visited infinitely often.

4. Let f be a function from the integers to the real numbers. Show that, for random walks on the line, the event

$$\sum_{i=1}^{\infty} f(n_i) < \infty$$

   is plausible, n_i being the position at time i.

5. Let $S = \sum_{i=1}^{\infty} \pm 2^{-i}$ be the series obtained by flipping a coin to decide whether a plus sign or a minus sign goes into the ith place. Show that the event $|S| < \varepsilon$ is plausible, and compute its probability. (*Hint:* See §1.2, exercise 7.)

6. Let X be a set, $\mathcal{F}$ a σ-field of subsets of X, and μ a probability measure of $\mathcal{F}$. Let $A_1, A_2, A_3, \ldots$ be a sequence of subsets of X belonging to $\mathcal{F}$.
   a. Show that, if $A_1 \supseteq A_2 \supseteq A_3 \cdots$, then

$$\mu\left(\bigcap_{i=1}^{\infty} A_i\right) = \lim_{i \to \infty} \mu(A_i)$$

b. Show that, if $A_1 \subseteq A_2 \subseteq A_3 \cdots$, then

$$\mu\left(\bigcup_{i=1}^{\infty} A_i\right) = \lim_{i \to \infty} \mu(A_i)$$

7. Let X be a set and $A_1, A_2, A_3, \ldots$ a sequence of subsets of X. The set

$$\bigcup_{k=1}^{\infty} \left(\bigcap_{n>k} A_n\right)$$

is denoted as $\liminf A_n$ or as $\{A_n; \text{a.a.}\}$ (abbreviations for *limes infimum* of A_n and "A_n, almost always," respectively). Show that $\{A_n^c; \text{a.a.}\}$ is the complement of $\{A_n; \text{i.o.}\}$. If X is a probability space and the A_n's are measurable, conclude that the probability of $\{A_n^c; \text{a.a.}\}$ is zero if and only if the probability of $\{A_n; \text{i.o.}\}$ is one.

8. Let X be a set, $\mathscr{F}$ a σ-field of subsets of X, and μ a probability measure on $\mathscr{F}$. Show that, if $A_1, A_2, A_3, \ldots$ are in $\mathscr{F}$, then

$$\mu(\liminf A_n) \leq \liminf \mu(A_n) \leq \limsup \mu(A_n) \leq \mu(\limsup A_n)$$

9. **a.** Let X be a set and let A_1 and A_2 be subsets of X. Show that the smallest σ-field containing A_1 and A_2 consists of at most 16 sets. (*Hint:* Take unions of the four sets: $A_1 \cap A_2, A_1^c \cap A_2, A_1 \cap A_2^c, A_1^c \cap A_2$.)

   b. Let $A_1, A_2, \ldots, A_k$ be subsets of X. Let $\mathscr{F}_k$ be the smallest σ-field containing the A_i's. Show that $\mathscr{F}_k$ has at most 2^{2^k} members.

   c. Show that the upper bound in part b cannot be improved. (*Hint:* Let M be the k-element set $\{p_1, \ldots, p_k\}$, and let $X = 2^M$ be the set of subsets of M. Let A_i be all subsets of M that contain the point p_i.

10. Let X be a set, $\mathscr{F}$ a σ-field of subsets of X, and μ a probability measure on $\mathscr{F}$. Suppose that $A_1, \ldots, A_n$ are independent sets belonging to $\mathscr{F}$.

   a. Show that $A_1^c, A_2, \ldots, A_n$ are independent.

   b. Let A be any one of the sets $A_1 \cap A_2,\ A_1^c \cap A_2,\ A_1 \cap A_2^c,\ A_1^c \cap A_2$. Show that $A, A_3, A_4, \ldots, A_n$ are independent.

   c. Let $\mathscr{F}_k$ be the smallest subfield of $\mathscr{F}$ containing $A_1, \ldots, A_k$. Show that if $A \in \mathscr{F}_k$ then $A, A_{k+1}, \ldots, A_n$ are independent.

   d. Let $\mathscr{F}_k$ be the smallest subfield of $\mathscr{F}$ containing $A_1, \ldots, A_k$ and $\mathscr{F}_{n-k}$ be the smallest subfield containing $A_{k+1}, \ldots, A_n$. Show that, if $A \in \mathscr{F}_k$ and $A' \in \mathscr{F}_{n-k}$, then A and A' are independent.

11. **a.** Let A_i be the subset of the unit interval corresponding to the event "H at the ith trial" in a Bernoulli sequence. Show that the A_i's are independent.

   b. Let B_i be the subset of the unit interval corresponding to the event "H T H at the ith, $i + 1$st and $i + 2$nd trial." Show that $B_1, B_4, B_7, B_{10}, \ldots$ are independent.

12. **a.** For the random walk with pauses, let A_i be the subset of the unit

interval corresponding to the event "a pause at time i." Show that the A_i's are independent.

b. For the random walk in the plane, let A_i be the subset of the unit interval corresponding to the event "an eastward move at time i." Show that the A_i's are independent.

13. Let N be a large positive integer. Prove that in a Bernoulli trial run lengths of length N occur infinitely often with probability one.

14. Prove that run lengths of *arbitrary* length occur infinitely often with probability one. (*Hint:* Consider the random pattern:

$$\mathrm{THTHHHTHHHT \ldots TH \ldots HT}$$

with the last term involving n H's. Let E_n be the event that this pattern occurs infinitely often and let E be the intersection of the events E_n.

15. For the random walk with pauses, prove that with probability one there are infinitely many pauses. Use the Borel–Cantelli lemma. (An alternative proof of this fact was suggested in §1.2, exercise 6.)

16. Let N be a large integer. Prove that the random walk on the line, starting at zero, passes either through the point N or the point $-N$ with probability one. Conclude that it passes through N with probability at least $\frac{1}{2}$.

17. Let $\mathbf{Z}^2$ be the integer points in the plane, and let Ω be a finite subset of $\mathbf{Z}^2$ containing the origin. Prove that a random path starting at the origin hits $\partial\Omega$ in a finite time with probability one.

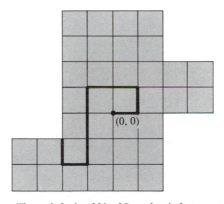

The path depicted hits $\partial\Omega$ on the ninth step.

18. In proving the law of large numbers in §1.1, we used Chebyshev's inequality to prove

$$\mu\left(\left\{\omega; \left|\frac{S_n(\omega)}{n}\right| < \varepsilon\right\}\right) \le 3n^{-2}\varepsilon^{-4}$$

Choose a sequence $\varepsilon_n \to 0$ such that $\sum_{n=1}^{\infty} n^{-2}\varepsilon_n^{-4}$ is finite and let A_n be the set

$$\left\{ \omega; \left| \frac{S_n(\omega)}{n} \right| > \varepsilon_n \right\}$$

Use the first Borel–Cantelli lemma to prove that $\mu(\{A_n; \text{i.o.}\}) = 0$ and deduce from this the law of large numbers.

19. Let $X = \{x_1, x_2, \ldots\}$ be a countable set, $P_1, P_2, \ldots$ a sequence of non-negative numbers such that $\sum P_i = 1$, and μ the measure

$$\mu(A) = \sum_{x_i \in A} P_i$$

Show that X cannot contain an infinite sequence of independent sets $A_1, A_2, \ldots$ such that, for all i, $\mu(A_i) = \frac{1}{2}$. (*Hint:* Start by observing that every point $x \in X$ must lie in one of the four sets $A_1 \cap A_2$, $A_1^c \cap A_2$, $A_1 \cap A_2^c$, or $A_1^c \cap A_2^c$. Thus the measure of the one-point set $\{x\}$ is less than or equal to $\frac{1}{4}$. Notice, by the way, the moral of this exercise: A *discrete* measure theoretic model for the Bernoulli process does not exist. Just let A_i be the subset of X corresponding to the event "an H at the ith trial.")

Chapter 2
Integration

Now that we have the tools of measure theory, we are ready to discuss integration. The student of the Riemann integral is accustomed to considering only integrals over subsets of $\mathbf{R}^n$. However, we will see that integrals can be defined whenever we have a triple $(X, \mathscr{F}, \mu)$, where X is a set, $\mathscr{F}$ is a σ-field of subsets of X, and μ is a measure defined on $\mathscr{F}$. Such a triple is called a *measure space*. Our basic example of a measure space is, of course, Lebesgue measure on the Borel sets of $[0, 1]$ or of the whole real line. The last section of Chapter 1 suggests that the theory of probability is rife with other examples. Notice that for the real line some sets have infinite measure— for instance, $\mu_L(\mathbf{R}) = \infty$. We will allow this to occur in general. (See the comments at the end of §1.3.)

§2.1 Measurable Functions

In the study of integration, it is convenient to allow functions to assume the values $+\infty$ and $-\infty$. To make this notion concrete, we define the *extended real* number system to be the set $\mathbf{R} \cup \{+\infty\} \cup \{-\infty\}$.

The elements $+\infty$ and $-\infty$ in the extended reals have the special properties

1. $-\infty < a < +\infty, \quad a \in \mathbf{R}$
2. $a + (\pm\infty) = \pm\infty, \quad a \in \mathbf{R}$
3. $a \cdot (\pm\infty) = \pm\infty, \quad a \in \mathbf{R}, a > 0$
4. $-1 \cdot (\pm\infty) = \mp\infty$

Now let $(X, \mathscr{F}, \mu)$ be a measure space. Let f be a function on X with values in the extended real numbers.

Definition 1. The function f is *measurable* if, for all $a \in \mathbf{R}$, the set $\{x \in X;\ f(x) > a\}$ is an element of $\mathscr{F}$.

Measurable functions can be characterized in various ways.

Proposition 2. The following are equivalent:

1. For all $a \in \mathbf{R}$, $\{x;\ f(x) > a\} \in \mathscr{F}$
2. For all $a \in \mathbf{R}$, $\{x;\ f(x) \geq a\} \in \mathscr{F}$
3. For all $a \in \mathbf{R}$, $\{x;\ f(x) < a\} \in \mathscr{F}$
4. For all $a \in \mathbf{R}$, $\{x;\ f(x) \leq a\} \in \mathscr{F}$

Proof.

$1 \Leftrightarrow 4$: The sets in items 1 and 4 are complementary and, because $\mathscr{F}$ is a
    σ-field, we know that $A \in \mathscr{F} \Leftrightarrow A^c \in \mathscr{F}$.

$2 \Leftrightarrow 3$: Same as above.

$1 \Rightarrow 2$: For all $a \in \mathbf{R}$

$$\{x;\ f(x) \geq a\} = \bigcap_{n=1}^{\infty} \left\{ x;\ f(x) > a - \frac{1}{n} \right\}$$

By item 1 each set $\{x;\ f(x) > a - 1/n\} \in \mathscr{F}$. Because $\mathscr{F}$ is a σ-field, the countable intersection is in $\mathscr{F}$.

$2 \Rightarrow 1$: For $a \in \mathbf{R}$

$$\{x;\ f(x) > a\} = \bigcup_{n=1}^{\infty} \left\{ x;\ f(x) \geq a + \frac{1}{n} \right\}$$

By item 2 each set $\{x;\ f(x) \geq a + 1/n\} \in \mathscr{F}$. Because $\mathscr{F}$ is a σ-field, the countable union is in $\mathscr{F}$. $\square$

In keeping with our notion of extended real numbers, we define the
extended Borel sets as the collection of subsets of $\mathbf{R} \cup \{+\infty\} \cup \{-\infty\}$ having
one of the following forms:

$$A, \quad A \cup \{+\infty\}, \quad A \cup \{-\infty\}, \quad A \cup \{+\infty, -\infty\}$$

where A is a Borel set. One can easily see that the extended Borel sets form a
σ-field.

Theorem 3. Conditions 1 through 4 of proposition 2 are equivalent to

5. For every extended Borel set B

(1) $$\{x;\ f(x) \in B\} \in \mathscr{F}$$

Proof. It is obvious that $5 \Rightarrow 1, 2, 3, 4$. We will show that $1, 2, 3, 4 \Rightarrow 5$.

Let $\mathscr{C}$ be the collection of all subsets C of $\mathbf{R} \cup \{+\infty\} \cup \{-\infty\}$ with the
property that

(2) $$\{x; f(x) \in C\} \in \mathscr{F}$$

We need to show that the extended Borel sets are contained in $\mathscr{C}$.

Note that, by items 1, 2, 3, and 4, the sets $(a, +\infty]$, $[a, +\infty]$, $[-\infty, a)$, and $[-\infty, a]$ are members of $\mathscr{C}$.

Also notice that $\mathscr{C}$ is a σ-field. Indeed, if $A_i \in \mathscr{C}$, $1 \le i < \infty$, then

$$\left\{x; f(x) \in \bigcap_{i=1}^{\infty} A_i\right\} = \bigcap_{i=1}^{\infty} \{x; f(x) \in A_i\} \in \mathscr{F}$$

so $\bigcap_{i=1}^{\infty} A_i \in \mathscr{C}$. Similarly, $\bigcup_{i=1}^{\infty} A_i \in \mathscr{C}$.

Now notice that the extended Borel sets are the smallest σ-field containing all the infinite intervals mentioned above. Thus the extended Borel sets must be contained in $\mathscr{C}$. $\square$

Example 4. Let $X = \mathbf{R}^n$, and let $\mathscr{F} = \mathscr{M}$ be the Lebesgue measurable sets. If $f : \mathbf{R}^n \to \mathbf{R}$ is continuous, then f is measurable. Indeed, if $a \in \mathbf{R}$, then $\{x \in \mathbf{R}^n; f(x) > a\}$ is open and thus is a Borel set.

Example 5. Let $X = \mathscr{B}$, the sample space for the Bernoulli process. As usual, identify $\mathscr{B}$ with I, the unit interval.

Consider $R_n(\omega)$, the Rademacher functions. These are measurable because they are piecewise constant; namely, for any subset A of $R \cup \{+\infty\} \cup \{-\infty\}$, we have that $\{\omega \in I; R_n(\omega) \in A\}$ is a finite union of intervals.

Similarly, if we let $S_n(\omega) = \sum_{k=1}^{n} R_k(\omega)$, we see that S_n is also piecewise constant and hence measurable.

Notation. In §1.2 we used the term *random variable* for a function $f : X \to \mathbf{R}$ when $(X, \mathscr{F}, \mu)$ is a probability space. The interesting functions on X are always measurable. For this reason we use the name *random variable* interchangeably with *measurable function* when discussing probabilistic notions.

Example 6. Let $T(\omega)$ be the number of times the random walk corresponding to ω returns to 0, and let $l_n(\omega)$ be the number of consecutive H's beginning at the nth toss of ω. As an exercise, show that T and l_n are random variables. (See exercise 4.)

Remark. The variables l_n and T are pathologically discontinuous. The possibility of integrating such functions, that is, computing their expectation values, vindicates the work we are about to put into the theory of integration. For example, we will show in Chapter 3 that $T = +\infty$ with probability one; that is, with probability one a random walk returns to 0 infinitely often.

We continue now with the study of the properties of measurable functions. Let $(X, \mathscr{F}, \mu)$ be a measure space.

Theorem 7. Let f and g be measurable functions on X. Let $\max(f,g)$ be the function on X defined by $\max(f,g)(x) = \max[f(x), g(x)]$. Then $\max(f,g)$ is measurable. Similarly, if $\min(f,g)(x) = \min[f(x), g(x)]$, then $\min(f,g)$ is measurable.

Proof. $\{x; \max(f,g)(x) > a\} = \{x; f(x) > a\} \cup \{x; g(x) > a\}$ and is thus an element of $\mathscr{F}$.

Similarly, $\{x; \min(f,g)(x) > a\} = \{x; f(x) > a\} \cap \{x; g(x) > a\}$ is also in $\mathscr{F}$. $\qquad\square$

Corollary 8. Let f be a measurable function on X. Let

$$f_+(x) = \begin{cases} f(x) & \text{if } f(x) \geq 0 \\ 0 & \text{if } f(x) < 0 \end{cases} \quad \text{and} \quad f_-(x) = \begin{cases} -f(x) & \text{if } f(x) \leq 0 \\ 0 & \text{if } f(x) > 0 \end{cases}$$

Then $f_+(x)$ and $f_-(x)$ are measurable.

Proof. $f_+(x) = \max(f, 0)$ and $f_-(x) = \max(-f, 0)$. $\qquad\square$

Notice that $f(x) = f_+(x) - f_-(x)$, so we have just shown that every measurable function can be written as the difference of two nonnegative measurable functions. (See the figure below.)

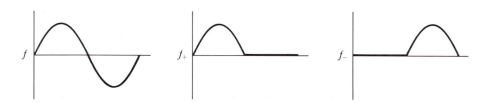

We now consider how measurable functions behave under limits. First, if $f_1, f_2, \ldots$ are functions on X, define functions $\sup f_i$ and $\inf f_i$ on X by

$$\sup_{i \geq 0} f_i(x) = \sup\{f_i(x); 1 \leq i < \infty\}$$

(3)

$$\inf_{i \geq 0} f_i(x) = \inf\{f_i(x); 1 \leq i < \infty\}$$

Theorem 9. If $f_1, f_2, \ldots$ is a sequence of measurable functions, then $\sup_{i > 0} f_i$ and $\inf_{i > 0} f_i$ are also measurable.

Proof. For $a \in \mathbf{R}$,

$$\{x; \sup f_i(x) > a\} = \bigcup_{i=1}^{\infty} \{x; f_i(x) > a\} \in \mathscr{F}$$

Similarly,

$$\{x;\ \inf f_i(x) \geq a\} = \bigcap_{i=1}^{\infty} \{x;\ f_i(x) \geq a\} \in \mathcal{F} \qquad \square$$

Now, if $f_1, f_2, \ldots$ is a sequence of functions, let $g_k = \sup_{n>k} f_n$. Notice that $g_1 \geq g_2 \geq g_3 \geq \cdots$ so that

$$\lim_{k \to \infty} g_k(x) = \inf_{k>0} g_k(x)$$

Recall that

$$\limsup f_i = \inf_{k>0} g_k$$

is called the "lim sup (*limes supremum*) of the sequence $f_1, f_2, \ldots$." Similarly,

$$\liminf f_i = \sup_{k>0} h_k \qquad \text{where} \qquad h_k = \inf_{n \geq k} f_n$$

Theorem 10. If $f_1, f_2, \ldots$ is a sequence of measurable functions on X, then $\limsup f_n$ and $\liminf f_n$ are measurable.

Proof. From the previous theorem, $g_k = \sup_{n \geq k} f_n$ and $h_k = \inf_{n \geq k} f_n$ are measurable. Applying theorem 9 once again, we have that

$$\limsup f_n = \inf_{k>0} g_k \qquad \text{and} \qquad \liminf f_n = \sup_{k>0} h_k$$

are also measurable. $\qquad \square$

Corollary 11. Let $f_1, f_2, \ldots$ be a sequence of measurable functions on X. Suppose that the f_n's converge pointwise to a function f; then f is also measurable.

Proof. $f = \limsup f_n = \liminf f_n$ $\qquad \square$
Thus we see that we can generate new examples of measurable functions by taking pointwise limits. Another way of getting new measurable functions is by composition.

Theorem 12. Let $f : X \to \mathbf{R}$ be a measurable function. Let g be a continuous function on $\mathbf{R}$. Then $g \circ f$ is measurable.

Proof. Note that we are not allowing f to take the values of $+\infty$ or $-\infty$ so that $g \circ f$ is defined.
Now, for all $a \in \mathbf{R}$, let $\mathcal{O}_a = \{t \in \mathbf{R};\ g(t) > a\}$. Then $\{x \in X;\ g \circ f(x) > a\} = \{x \in X;\ f(x) \in \mathcal{O}_a\}$. Because g is continuous, $\mathcal{O}_a$ is a Borel set. So $\{x \in X;\ f(x) \in \mathcal{O}_a\} \in \mathcal{F}$, because f is measurable. $\qquad \square$

Example 13. If $f : X \to \mathbf{R}$ is measurable, then $e^{i\lambda f}, e^{-f^2/2}, \sin f, |f|$, and so on are measurable.

In a similar vein we have the following theorem.

Theorem 14. Let $f_1, f_2, \ldots, f_n$ be measurable functions on X with $f_i : X \to \mathbf{R}$, $1 \le i \le n$. Let $G : \mathbf{R}^n \to \mathbf{R}$ be continuous. Then $G(f_1, \ldots, f_n)$ is measurable.

We leave the proof to the reader as an exercise.

Example 15. Let $f_i : X \to \mathbf{R}$, $i = 1, 2$, be measurable functions. Since $x + y$ and xy are continuous functions from $\mathbf{R}^2$ to $\mathbf{R}$, $f_1 + f_2$ and $f_1 f_2$ are measurable.

Exercises for §2.1

1. Given a set X and subsets $A_1, A_2, \ldots$, let

$$A_+ = \limsup A_n = \{A_n; \text{i.o.}\} = \bigcap_{k=1}^{\infty} \bigcup_{n \ge k} A_n$$

and

$$A_- = \liminf A_n = \{A_n; \text{a.a.}\} = \bigcup_{k=1}^{\infty} \bigcap_{n \ge k} A_n$$

Let f_+, f_- and f_n be the characteristic functions of the set A_+, A_- and A_n, respectively. Prove that $f_+ = \limsup f_n$ and $f_- = \liminf f_n$.

2. **a.** Let X be a set, $\mathscr{F}$ a σ-field of subsets of X, and μ a measure on $\mathscr{F}$. A map $f = (f_1, \ldots, f_n)$ of X into $\mathbf{R}^n$ is said to be *measurable* if each of the coordinate functions, f_i, is measurable. Show that f is measurable if and only if $f^{-1}(B) \in \mathscr{F}$ for every Borel set $B \subseteq \mathbf{R}^n$.

   b. Let g be a real-valued function on $\mathbf{R}^n$. Then g is *Borel measurable* if, for all numbers a, the set $\{x \in \mathbf{R}^n; g(x) > a\}$ is a Borel set. Prove that, if $f : X \to \mathbf{R}^n$ is measurable and $g : \mathbf{R}^n \to \mathbf{R}$ is Borel measurable, then $g \circ f : X \to \mathbf{R}$ is measurable.

3. **a.** Let $f_n : X \to \mathbf{R}$, $n = 1, 2, \ldots$, be a sequence of measurable functions. Show that the set of points, $x \in X$, where the sequence $\{f_n(x); n = 1, 2, \ldots\}$ converges, is measurable.

   b. From part a deduce that the set of points, $\omega \in I$, for which the "randomized harmonic series"

$$\sum_{n=1}^{\infty} \frac{R_n(\omega)}{n}$$

converges, is a measurable subset of I.

4. The examples below describe some random variables that occur naturally in the theory of Bernoulli trials. Show that these random variables are measurable functions on the unit interval I. (As usual we make the identification $\mathscr{B}$ = Bernoulli sequences = I.)

   a. R = ruin time = the time it takes a gambler with an initial stake of N dollars to be reduced to penury.

   b. l_n = the number of successive heads that appear beginning with the nth toss.

   c. T = the number of times a random walk returns to the origin.

   d. $\limsup S_n/n$, where S_n is the sum of the first n Rademacher functions. This random variable measures "violation of the law of large numbers."

5. Let T be the random variable in part c of exercise 4. Show that the set

$$\{\omega \in I;\ T(\omega) < \infty\}$$

   is uncountable and dense in I. (We will show in Chapter 3 that this set is of measure zero.)

6. **a.** Let S be the random variable in part d of exercise 4. Show that, for every subinterval of the unit interval and for every real number a with $-1 < a < 1$, $S(\omega) = a$ uncountably often. (*Hint:* Suppose that $\omega = .a_1 a_2 a_3 \ldots$ is a sequence for which $S(\omega) = a$. Let

$$\omega' = a_1\, b_1\, a_2\, a_3\, b_2\, a_4\, a_5\, a_6\, b_3\, a_7\, a_8\, a_9\, a_{10}\, b_4 \ldots$$

   Show that $S(\omega') = a$ no matter what the b_i's are.)

   b. On the other hand, show that the set $\{\omega \in I;\ S(\omega) \neq 0\}$ is of measure zero.

7. **a.** Let Ω be a finite subset of $\mathbf{Z}^2$ containing $(0, 0)$ as an interior point. (See §1.2, exercise 8.) Let H be the time at which a random walk starting at the origin hits $\partial\Omega$. (H is called the "hitting time.") Show that H, regarded as a function on the unit interval, is measurable and is finite except on a set of measure zero. (*Hint:* See §1.4, exercise 17.)

   b. Assume that the points $(1, 0)$ and $(0, 1)$ are also interior points. Show that the set

$$\{\omega \in I;\ H(\omega) = +\infty\}$$

   is uncountable.

8. Let $f : \mathbf{R} \to \mathbf{R}$ be monotone increasing. Show that f is measurable.

9. A function $f : \mathbf{R} \to \mathbf{R}$ is upper-semicontinuous at x if for every $\varepsilon > 0$ there exists a $\delta > 0$ such that $f(y) < f(x) + \varepsilon$ when $|x - y| < \delta$. Show that if f is upper-semicontinuous at all points of $\mathbf{R}$, it is measurable.

10. Two functions f and g are said to be equal almost everywhere if $f = g$ except on a set of measure zero. Show that if $f = g$ almost everywhere and if f is measurable, g is also measurable.

11. Let $(X, \mathcal{F}, \mu)$ be a measure space with $\mu(X) < \infty$. A sequence of measurable functions $f_n : X \to \mathbf{R}$, $n = 1, 2, \ldots$, is said to converge to zero *in measure* if for all $\varepsilon > 0$

$$\lim_{n \to \infty} \mu\{x \in X \,;\, |f_n(x)| > \varepsilon\} = 0$$

a. Prove that, if f_n converges to zero pointwise except on a set of measure zero, then f_n converges to zero in measure.

b. Show that the converse of part a is not true. (*Hint:* Let $X = I$, $\mathcal{F}$ be the Borel subsets of I, and μ be Lebesgue measure. Let $A_1 = [0, \frac{1}{2}]$, $A_2 = [\frac{1}{2}, 1]$, $A_3 = [0, \frac{1}{4}]$, $A_4 = [\frac{1}{4}, \frac{2}{4}]$, $A_5 = [\frac{2}{4}, \frac{3}{4}]$, $A_6 = [\frac{3}{4}, 1]$, $A_7 = [0, \frac{1}{8}]$, $A_8 = [\frac{1}{8}, \frac{2}{8}]$, and so on. Let f_n be the function that is 1 on A_n and 0 elsewhere.)

§2.2 The Lebesgue Integral

Let $(X, \mathcal{F}, \mu)$ be a measure space and $s : X \to \mathbf{R}$ be a measurable function. The function s is called a *simple function* if it takes on only a finite number of values.

Example 1. Let $E \in \mathcal{F}$ and define

$$1_E(x) = \begin{cases} 1 & \text{if } x \in E \\ 0 & \text{if } x \notin E \end{cases}$$

Then $1_E(x)$ is called the *characteristic function* of E; it is clearly simple.

In general, if s is a simple function taking on the values $c_1, \ldots, c_N$, let

$$E_i = s^{-1}(c_i) = \{x \in X \,;\, s(x) = c_i\} \qquad \{1 \le i \le N\}$$

We then have that

$$s(x) = \sum_{i=1}^{N} c_i 1_{E_i}(x)$$

It is easy to define the integral of nonnegative simple functions.

Definition 2. Let $s : X \to \mathbf{R}$ be a nonnegative simple function and let $E \in \mathcal{F}$. Let $c_1, \ldots, c_N$ be the distinct *nonzero* values of s and let $E_i = s^{-1}(c_i), 1 \le i \le N$. Define the *integral* of s over E with respect to μ as the sum

(1)
$$I_E(s) = \sum_{i=1}^{N} c_i \mu(E \cap E_i)$$

Remark. This value may be $+\infty$ because $\mu(E \cap E_i)$ can be $+\infty$.

Proposition 3. Let $s_i : X \to \mathbf{R}$, $i = 1, 2$, be nonnegative simple functions and let $E \in \mathscr{F}$.

1. (linearity)
   a. $I_E(cs_1) = cI_E(s_1)$ for $c \in \mathbf{R}$, $c \geq 0$
   b. $I_E(s_1 + s_2) = I_E(s_1) + I_E(s_2)$
2. (monotonicity) If $s_1 \leq s_2$ then $I_E(s_1) \leq I_E(s_2)$

Proof. 1a is clear.

1b: Let $c_1, \ldots, c_m$ be the distinct values of s_1 and $d_1, \ldots, d_n$ be the distinct values of s_2. Let $E_i = s_1^{-1}(c_i)$, $1 \leq i \leq m$, and $F_j = s_2^{-1}(d_j)$, $1 \leq j \leq n$. The E_i's form a mutually disjoint cover of X and so do the F_j's. Thus $E_i \cap F_j$, $1 \leq i \leq m$ and $1 \leq j \leq n$, also form a mutually disjoint cover of X, and $s_1 + s_2$ has the constant value $c_i + d_j$ on $E_i \cap F_j$. Hence

$$I_E(s_1 + s_2) = \sum_{i,j} (c_i + d_j)\mu(E_i \cap F_j \cap E)$$

$$= \sum_i c_i \sum_j \mu(E_i \cap F_j \cap E) + \sum_j d_j \sum_i \mu(E_i \cap F_j \cap E)$$

$$= \sum_i c_i \mu(E_i \cap E) + \sum_j d_j \mu(E_j \cap E)$$

$$= I_E(s_1) + I_E(s_2)$$

2: $s_2 - s_1$ is a nonnegative simple function, and $s_2 = s_1 + s_2 - s_1$, so $I(s_2) = I(s_1) + I(s_2 - s_1)$. $\square$

We now extend our notion of integration to nonnegative measurable functions by approximation with simple functions.

Definition 4. Let f be a nonnegative measurable function from X into the (nonnegative) extended real numbers and let $E \in \mathscr{F}$. Then the integral of f on E with respect to μ is defined by

(2) $$\int_E f \, d\mu = \sup\{I_E(s); 0 \leq s \leq f, s \text{ simple}\}$$

The following proposition shows that this definition of the integral is consistent with our previous definition when f is a simple function.

Proposition 5. $I_E(s) = \int_E s \, d\mu$ if s is a nonnegative simple function.

Proof. Clearly $I_E(s) \leq \int_E s \, d\mu$ because s is a simple function with $s \leq s$. To show equality let s' be any simple function with $0 \leq s' \leq s$. By monotonicity $I_E(s') \leq I_E(s)$. Hence $\sup\{I_E(s'); 0 \leq s' \leq s, s' \text{ simple}\} \leq I_E(s)$. $\square$

At this point the reader is probably asking why the integral of nonnegative measurable functions should be approximated by the integrals of simple functions. We justify this by showing that nonnegative measurable functions can themselves be approximated by nonnegative simple functions.

Theorem 6.   Let f be a nonnegative measurable function on X with values in the (nonnegative) extended real numbers. There exists a sequence of nonnegative simple functions

$$0 \leq s_1 \leq s_2 \leq \cdots \leq f$$

such that $s_i \to f$ pointwise. Moreover, if f is bounded, then $s_i \to f$ uniformly.

Proof.   We begin by defining s_n. Consider the interval $[0, n)$ on **R**. Divide this interval into $n2^n$ subintervals of length $1/2^n$; namely, let

$$(3) \qquad I_i = \left\{ t \in \mathbf{R}; \frac{i-1}{2^n} \leq t < \frac{i}{2^n} \right\}, \ 1 \leq i \leq n2^n$$

Let $E_i = f^{-1}(I_i)$ and $F_n = f^{-1}([n, +\infty])$. Together the E_i's with F_n form a mutually disjoint cover of X (n is fixed).
    Define

$$(4) \qquad s_n(x) = \sum_{i=1}^{n2^n} \left(\frac{i-1}{2^n} \right) 1_{E_i}(x) + n1_{F_n}(x)$$

Notice that on E_i we have

$$\frac{i-1}{2^n} \leq f \leq \frac{i}{2^n} \qquad \text{and} \qquad s_n = \frac{i-1}{2^n}$$

Thus, $s_n(x) \leq f(x)$ for $x \in E_i, \ i = 1, \ldots, n2^n$

Similarly, on F_n

$$n \leq f \qquad \text{and} \qquad s_n = n$$

so

$$s_n(x) \leq f(x) \qquad \text{for } x \in F_n$$

Hence

$$s_n \leq f \qquad \text{on all of } X$$

Notice also that $s_n \leq s_{n+1}$. Indeed, let I be one of the intervals $\left[\frac{i-1}{2^n}, \frac{i}{2^n} \right)$.
Notice that $I = I' \cup I''$ where

$$I' = \left[\frac{2i-2}{2^{n+1}}, \frac{2i-1}{2^{n+1}}\right) \quad \text{and} \quad I'' = \left[\frac{2i-1}{2^{n+1}}, \frac{2i}{2^{n+1}}\right)$$

Let $E = f^{-1}(I)$, $E' = f^{-1}(I')$, and $E'' = f^{-1}(I'')$. Then

$$s_n(x) = \frac{i-1}{2^n} \quad \text{for } x \in E$$

whereas

$$s_{n+1}(x) = \frac{i-1}{2^n} \quad \text{for } x \in E' \quad \text{and} \quad s_{n+1}(x) = \frac{2i-1}{2^{n+1}} \quad \text{for } x \in E''$$

Then, because

$$E = E' \cup E'' \quad \text{and} \quad \frac{i-1}{2^n} < \frac{2i-1}{2^{n+1}}$$

we have shown that

$$s_n(x) \le s_{n+1}(x) \quad \text{for all } x \in E$$

This argument is clearly independent of which I_i we began with. It also works with minor changes on $[n, +\infty]$, so

$$s_n(x) \le s_{n+1}(x) \quad \text{for all } x \in X$$

We now show that $s_n \to f$ pointwise. Two separate cases are involved.
Case 1. $f(x) = +\infty$
  In this case $x \in F_n$ for all n, so $s_n(x) = n$ for all n. That is, $s_n(x) \to +\infty$.
Case 2. $f(x)$ is finite
  Say $f(x) < n_0$. Then, for all $n > n_0$, $f(x)$ lies on one of the intervals I_i; that is,

$$\frac{i-1}{2^n} \le f(x) < \frac{i}{2^n}$$

But then

$$s_n(x) = \frac{i-1}{2^n} \quad \text{so} \quad |f(x) - s_n(x)| < \frac{1}{2^n}$$

for $n > n_0$, proving that $s_n(x) \to f(x)$.
  Finally, suppose that f is bounded, say $f(x) < n_0$ for all $x \in X$. The preceding argument shows that for $n > n_0$

$$|f(x) - s_n(x)| < \frac{1}{2^n} \quad \text{for all } x \in X$$

That is, $s_n \to f$ uniformly on X. $\qquad\qquad\qquad\qquad\qquad\qquad\qquad \square$

Remark. Hidden in this proof is a crucial ingredient of the Lebesgue theory. In the Rıemann theory one approximates a function f by simple functions by dividing the *domain* of the function into small intervals, as shown in part (a) of the figure. In the Lebesgue theory one approximates by dividing the *range* of the function into small intervals, as shown in part b of the figure.

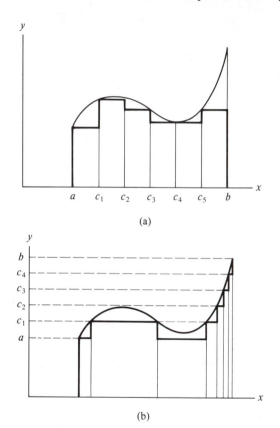

Graph of f. (The dark lines represent the approximating function.)

The second procedure has two conspicuous advantages. First, the x axis no longer has to be the real line; it can be any measure space X. Second, one gets a good approximation of f by simple functions without assuming f to be continuous.

We now look at some properties of the integral.

Proposition 7. Let E, $F \in \mathscr{F}$ and let f and g be nonnegative measurable functions.

1. (monotonicity) If $f \leq g$ then

$$\int_E f \, d\mu \leq \int_E g \, d\mu$$

2. If $E \subset F$ then

$$\int_E f \, d\mu \leq \int_F f \, d\mu$$

3. If $\mu(E) = 0$ then

$$\int_E f \, d\mu = 0$$

Proof.

1. This is obvious because, if s is a simple function and $s \leq f$, then $s \leq g$.
2. We first verify this for

$$f = 1_G \qquad \text{where } G \in \mathscr{F}$$

Then $\int_E (1_G) \, d\mu = I_E(1_G) = \mu(E \cap G)$ and likewise $\int_F (1_G) \, d\mu = \mu(F \cap G)$. But $E \cap G \subseteq F \cap G$, hence

$$\mu(E \cap G) \leq \mu(F \cap G)$$

Now, by the linearity of I_E we know that $\int_E s \, d\mu \leq \int_F s \, d\mu$ when s is a simple function. But, by the definition of $\int_E f \, d\mu$ and $\int_F f \, d\mu$, the statement has to be true in general.
3. If $f = s$ is a simple function, this assertion is clear; thus

$$\sup\{I_E(s); \, 0 \leq s \leq f\} = 0 \qquad \qquad \square$$

Remark. We will defer to §2.3 the proof that, for nonnegative measurable functions,

$$\int_E (f_1 + f_2) \, d\mu = \int_E f_1 \, d\mu + \int_E f_2 \, d\mu$$

Unfortunately, this fact, which looks as though it should be practically obvious, requires a somewhat delicate proof.

We can also now prove Chebyshev's inequality in full generality.

Theorem 8. (Chebyshev) Let f be a nonnegative measurable function. For $E \in \mathscr{F}$ and $c > 0$, let $E_c = \{x \in E; f(x) \geq c\}$. Then

(5)
$$\mu(E_c) \leq \frac{1}{c} \int_E f \, d\mu$$

Proof. Because $f \geq c$ on E_c,

$$\int_{E_c} c \, d\mu \leq \int_{E_c} f \, d\mu \qquad \text{(by monotonicity)}$$

But

$$\int_{E_c} c \, d\mu = I_{E_c}(c) = c\mu(E_c)$$

Thus

$$c\mu(E_c) \leq \int_{E_c} f \, d\mu \leq \int_E f \, d\mu \qquad \qquad \square$$

Corollary 9. If f is a nonnegative measurable function with $\int_E f \, d\mu < \infty$, then $\{x \in E; f(x) = +\infty\}$ has measure zero.

Notation. If a property holds on a set $E \in \mathcal{F}$ except for a subset of measure zero, we say that the property holds *almost everywhere* on E (abbreviated as a.e.). Thus corollary 9 can be restated as

(6) $$\int_E f \, d\mu < \infty \Rightarrow f(x) < \infty \qquad \text{a.e. on } E$$

Proof. Let $A_n = \{x \in E; f(x) \geq n\}$, and let $A = \{x \in E; f(x) = +\infty\}$. By Chebyshev

$$\mu(A_n) \leq \frac{1}{n} \int_E f \, d\mu$$

But $A \subset A_n$ for all n, so

$$\mu(A) \leq \mu(A_n) \leq \frac{1}{n} \int_E f \, d\mu \qquad \text{for all } n$$

Because $\int_E f \, d\mu < \infty$ we must have $\mu(A) = 0$. $\square$

Corollary 10. Let f be a nonnegative measurable function and let $E \in \mathcal{F}$. If $\int_E f \, d\mu = 0$, then $f = 0$ a.e. on E.

Proof. Let $A = \{x \in E; f(x) \neq 0\}$, and let $A_n = \{x \in E; f(x) > 1/n\}$. Because $A = \bigcup_{n=1}^{\infty} A_n$ it is enough to show that $\mu(A_n) = 0$. By Chebyshev

$$\mu(A_n) \leq n \int_E f \, d\mu = 0 \qquad \qquad \square$$

Another property we can now prove is that the integral behaves nicely on disjoint unions of sets.

Theorem 11.   Let f be a nonnegative measurable function on X. Let $A_1, A_2,$ ... be a sequence of pairwise disjoint members of $\mathscr{F}$. Let $A = \bigcup_{i=1}^{\infty} A_i$. Then

(7)
$$\int_A f \, d\mu = \sum_{i=1}^{\infty} \int_{A_i} f \, d\mu$$

Proof.   First we prove the theorem in the case that $f(x) = 1_E(x)$ for some $E \in \mathscr{F}$. In this case

$$\int_A 1_E \, d\mu = \mu(A \cap E) \qquad \text{and} \qquad \int_{A_i} 1_E \, d\mu = \mu(A_i \cap E)$$

By countable additivity of μ, we know that

$$\mu(A \cap E) = \sum_{i=1}^{\infty} \mu(A_i \cap E)$$

That is,

$$\int_A 1_E \, d\mu = \sum_{i=1}^{\infty} \int_{A_i} 1_E \, d\mu$$

By the linearity of I_A and I_{A_i}, the theorem is now automatically true for all simple functions. We prove the general case with f being an arbitrary non-negative measurable function as follows.

For $\varepsilon > 0$ pick a simple function $s \leq f$ with

$$\int_A f \, d\mu \leq I_A(s) + \varepsilon$$

Because the theorem is true for simple functions, we have that

$$I_A(s) = \sum_{i=1}^{\infty} I_{A_i}(s) \leq \sum_{i=1}^{\infty} \int_{A_i} f \, d\mu$$

so
$$\int_A f \, d\mu \leq \sum_{i=1}^{\infty} \int_{A_i} f \, d\mu + \varepsilon$$

Because ε is arbitrary, we must have

(8)
$$\int_A f \, d\mu \leq \sum_{i=1}^{\infty} \int_{A_i} f \, d\mu$$

To prove the opposite inequality, first consider two disjoint sets $A_1, A_2 \in \mathscr{F}$. Let s_1 and s_2 be simple functions with $0 \leq s_i \leq f$ and

(9)
$$\int_{A_i} s_i \, d\mu \geq \int_{A_i} f \, d\mu - \frac{\varepsilon}{2} \qquad \text{for } i = 1, 2$$

Let $s = \max(s_1, s_2)$; then $0 \leq s \leq f$, and s is a simple function. Also $s_1 \leq s$ and $s_2 \leq s$, so we can replace s_i by s in inequality 9; namely,

$$\int_{A_i} s \, d\mu \geq \int_{A_i} f \, d\mu - \frac{\varepsilon}{2} \qquad \text{for } i = 1, 2$$

Adding these inequalities gives

$$\int_{A_1} s \, d\mu + \int_{A_2} s \, d\mu \geq \int_{A_1} f \, d\mu + \int_{A_2} f \, d\mu - \varepsilon$$

or

$$\int_{A} s \, d\mu \geq \int_{A_1} f \, d\mu + \int_{A_2} f \, d\mu - \varepsilon$$

because the theorem is known to be true for simple functions.

Now $\int_A f \, d\mu \geq \int_A s \, d\mu$, so

$$\int_{A} f \, d\mu \geq \int_{A_1} f \, d\mu + \int_{A_2} f \, d\mu - \varepsilon$$

Because ε is arbitrary,

$$\int_{A} f \, d\mu \geq \int_{A_1} f \, d\mu + \int_{A_2} f \, d\mu$$

Thus we have the inequality for A_1 and A_2. An induction argument gives

(10)
$$\int_{A_1 \cup A_2 \cup \cdots \cup A_n} f \, d\mu \geq \sum_{i=1}^{n} \int_{A_i} f \, d\mu$$

Returning to the general situation—that is, $A = \bigcup_{i=1}^{\infty} A_i$—we have

$$\int_{A} f \, d\mu \geq \int_{A_1 \cup A_2 \cup \cdots \cup A_n} f \, d\mu$$

because $A_1 \cup A_2 \cup \cdots \cup A_n \subset A$. Hence, by inequality 10,

$$\int_{A} f \, d\mu \geq \sum_{i=1}^{n} \int_{A_i} f \, d\mu \qquad \text{for all } n$$

That is,

$$\int_{A} f \, d\mu \geq \sum_{i=1}^{\infty} \int_{A_i} f \, d\mu \qquad \qquad \square$$

Application.   Theorem 11 tells us that we can use the integral to define new measures. For example, we define *Gaussian measure*, μ_G, on the measurable subsets of **R** by

(11)
$$\mu_G(A) = \frac{1}{\sqrt{2\pi}} \int_A e^{-x^2/2} \, d\mu_L$$

Theorem 11 says that this is countably additive and so is indeed a measure. In fact, μ_G is a probability measure because

$$\frac{1}{\sqrt{2\pi}} \int_{\mathbf{R}} e^{-x^2/2} \, d\mu_L = 1$$

The theorem also has the following corollary.

Corollary 12. Suppose f and g are nonnegative measurable functions and $E \in \mathscr{F}$. Then, if $f = g$ a.e. on E,

$$\int_E f \, d\mu = \int_E g \, d\mu$$

Proof. Let $A_1 = \{x \in E; \, f(x) = g(x)\}$ and $A_2 = \{x \in E; \, f(x) \neq g(x)\}$. Clearly A_1 and A_2 are disjoint. Moreover, by assumption $\mu(A_2) = 0$, so

$$\int_{A_2} f \, d\mu = \int_{A_2} g \, d\mu = 0$$

Also $\int_{A_1} f \, d\mu = \int_{A_1} g \, d\mu$ because $f = g$ on A_1. Thus

$$\int_E f \, d\mu = \int_{A_1} f \, d\mu + \int_{A_2} f \, d\mu = \int_{A_1} g \, d\mu + \int_{A_2} g \, d\mu = \int_E g \, d\mu \qquad \square$$

Exercises for §2.2

1. Let I be the unit interval. Show that

$$\int_I x \, d\mu_L = \frac{1}{2}$$

using only properties of the Lebesgue integral discussed in this section. (*Hint:* Approximate x from above and from below by simple functions.)

2. Let J be the interval $1 \leq x < \infty$. Show that

$$\int_J \left(\frac{1}{x}\right) d\mu_L = +\infty$$

using only properties of the Lebesgue integral discussed in this section.

3. Let $(X, \mathscr{F}, \mu)$ be a probability space. Let $f : X \to [0, +\infty)$ be a random variable (that is, measurable function). The integral

$$E = \int_X f \, d\mu$$

is called the expectation value of f (or "most likely value" of f) and the integral

$$V = \int_X (f - E)^2 \, d\mu$$

is called the *variance* of f. Show that, if the variance of f is small, f deviates from its expectation value with very small probability. Explicitly, show that the probability that f deviates by ε from E—that is,

$$\mu(\{x \in X ; |f(x) - E| > \varepsilon\})$$

is less than or equal to $\varepsilon^{-2} V$.

4. Let H_n be the number of heads occurring in the first n trials of a Bernoulli sequence. Compute its expectation value and variance.

5. Consider the "random" series

$$1 \pm \frac{1}{2} \pm \frac{1}{4} \pm \frac{1}{8} \pm \cdots$$

with the assignment of a plus or minus in the nth term being decided by the toss of a coin. Compute its expectation value and variance.

6. Let $(X, \mathscr{F}, \mu)$ be a measure space and f a nonnegative measurable function. For all $a \in (0, \infty)$, let

$$\Phi(a) = \mu(\{x \in X ; f(x) > a\})$$

Suppose $\int_X f^k \, d\mu < \infty$, $k > 0$. Show that there exists a constant $C > 0$ such that

$$\Phi(a) \leq C a^{-k}$$

That is, show that Φ goes to zero at least as fast as a^{-k} as $a \to +\infty$.

7. Let J be a finite subinterval of the real line and $f : J \to \mathbf{R}$ a simple function taking on values $c_1, \ldots, c_n$. The function f is called a *step function* if $f^{-1}(c_i)$ is a finite union of intervals for each i. Given a simple function $s : J \to \mathbf{R}$ and a positive number ε, show that there exists a step function f such that

(*) $$\int_J |s - f| \, d\mu_L < \varepsilon$$

(*Hint:* Show that, if A is a measurable subset of J, there exists a finite union of intervals B such that $d(A, B) = \mu(S(A, B)) < \varepsilon$. Now prove the inequality (∗) for $s = 1_A$. Proceed.)

8. Let X be a countable set; that is,

$$X = \{x_1, x_2, x_3, \ldots\}$$

and let $P_1, P_2, P_3, \ldots$ be a sequence of nonnegative numbers such that $\sum P_i = 1$. For $A \subseteq X$ let

$$\mu(A) = \sum_{x_i \in A} P_i$$

We saw in §1.4 that μ is a probability measure on the σ-field of all subsets of X. Prove that every function $f : X \to \mathbf{R} \cup \{\pm\infty\}$ is measurable, and prove that, for f nonnegative,

$$\int_X f \, d\mu = \sum P_i f(x_i)$$

9. Let $f : \mathbf{R} \to [0, \infty)$ be measurable. Given $a \in \mathbf{R}$, let $f_a(x) = f(x - a)$. Show that f_a is measurable and that

$$\int f_a \, d\mu_L = \int f \, d\mu_L$$

(*Hint:* See §1.3, exercise 14.)

10. Let $(X, \mathscr{F}, \mu)$ be a measure space and A and B be measurable subsets of X. Show that, if $\mu(S(A, B)) = 0$, then, for every nonnegative measurable function f,

$$\int_A f \, d\mu = \int_B f \, d\mu$$

11. Let $(X, \mathscr{F}, \mu)$ be a measure space and f and g positive measurable functions. Show that, if g is simple, then, for all $E \in \mathscr{F}$,

$$\int_E (f + g) \, d\mu = \int_E f \, d\mu + \int_E g \, d\mu$$

12. Let $(X, \mathscr{F}, \mu)$ be a measure space with $\mu(X) < \infty$. Let f be a bounded nonnegative measurable function and let $\{s_n\}$ be the sequence of simple functions constructed in theorem 6. Show that, for $E \in \mathscr{F}$,

$$\int_E s_n \, d\mu \to \int_E f \, d\mu$$

(*Hint:* The sequence s_n converges uniformly to f. Moreover,

$$\int_E (f - s_n)\, d\mu + \int_E s_n\, d\mu = \int_E f\, d\mu$$

by exercise 11.)

§2.3 Further Properties of the Integral; Convergence Theorems

The reader has probably noticed that the definition of the integral does not make evaluation of integrals easy. The set of simple functions s with $0 \le s \le f$ is a formidably large set. In this section we will develop some effective techniques for computing integrals and for manipulating integrals and limits. The three key results are the monotone convergence theorem, Fatou's lemma, and the Lebesgue dominated convergence theorem.

Let $f_1, f_2, \ldots$ be a sequence of measurable functions with

$$0 \le f_1 \le f_2 \le \cdots$$

Note that $f = \lim_{i \to \infty} f_i$ exists and is measurable.

Theorem 1. (Monotone convergence theorem) Let f and f_i, $i = 1, 2, 3, \ldots$, be as above. Then

(1) $$\int_E f\, d\mu = \lim_{i \to \infty} \int_E f_i\, d\mu \qquad \text{for } E \in \mathcal{F}$$

To prove this theorem we need the following lemma.

Lemma 2. Let f be a nonnegative measurable function on X and let $E_1, E_2, E_3, \ldots$ be a sequence of sets in $\mathcal{F}$ with $E_1 \subset E_2 \subset E_3 \subset \cdots$. Set $E = \bigcup_{i=1}^{\infty} E_i$; then

$$\int_E f\, d\mu = \lim_{i \to \infty} \int_{E_i} f\, d\mu$$

Proof. Let
$$A_1 = E_1$$
$$A_2 = E_2 - E_1$$
$$A_3 = E_3 - E_2$$

and so on. Then the A_i's are pairwise disjoint and

$$\bigcup_{i=1}^{\infty} A_i = E \qquad \text{and} \qquad \bigcup_{i=1}^{n} A_i = E_n$$

So by countable additivity

$$\int_E f\,d\mu = \sum_{i=1}^{\infty} \int_{A_i} f\,d\mu$$

$$= \lim_{n\to\infty} \sum_{i=1}^{n} \int_{A_i} f\,d\mu$$

$$= \lim_{n\to\infty} \int_{E_n} f\,d\mu \qquad\qquad \triangledown$$

We now prove the monotone convergence theorem. We have

$$0 \le f_1 \le f_2 \le \cdots \le f = \lim_{n\to\infty} f_n$$

By monotonicity

$$\int_E f_1\,d\mu \le \int_E f_2\,d\mu \le \cdots \le \int_E f\,d\mu$$

so $\lim_{n\to\infty} \int_E f_n\,d\mu$ exists and must be less than or equal to $\int_E f\,d\mu$. Let $a = \lim_{n\to\infty} \int_E f_n\,d\mu$. We must establish that

(2)
$$a \ge \int_E f\,d\mu$$

Let s be any simple function, with $0 \le s \le f$, and let $c \in \mathbf{R}$, with $0 < c < 1$. Let

$$E_n = \{x \in E;\ f_n(x) \ge cs(x)\}$$

Notice that $E_1 \subset E_2 \subset \cdots$ because $f_1 \le f_2 \le \cdots$. Also notice that $\bigcup_{n=1}^{\infty} E_n = E$. Indeed, if $x \in E$ with $s(x) = 0$, then $x \in E_n$ for all n, and, if $x \in E$ with $s(x) \ne 0$, then because $c < 1$

$$f(x) \ge s(x) > cs(x)$$

So, for some n, $f_n(x) \ge cs(x)$ because

$$f_n(x) \to f(x)$$

that is, $x \in E_n$.

Taking integrals, we get

$$a = \lim_{n\to\infty} \int_E f_n\,d\mu \ge \int_E f_n\,d\mu \ge \int_{E_n} f_n\,d\mu \ge \int_{E_n} cs\,d\mu$$

because $f_n \ge cs$ on E_n. We apply the lemma to $E = \bigcup_{n=1}^{\infty} E_n$ to get

$$a \geq \lim_{n \to \infty} \int_{E_n} cs(x)\, d\mu = \int_E cs(x)\, d\mu = c \int_E s(x)\, d\mu$$

Because this is true for all c with $0 < c < 1$, it must also be true that

$$a \geq \int_E s\, d\mu$$

By now taking the supremum over all simple s with $0 \leq s \leq f$, we get inequality 2.   □

Remark. Thanks to the monotone convergence theorem, we can apply theorem 6 of §2.2 to the evaluation of integrals. If f is a nonnegative measurable function and s_n is the sequence of simple functions constructed in theorem 6 of §2.2, then

(3) $$\int_E s_n\, d\mu \to \int_E f\, d\mu$$

We will use this formula to clear up some matters that we left dangling in §2.2.

Theorem 3. Let f and g be nonnegative measurable functions and let $c > 0$ be a real number. For $E \in \mathscr{F}$ we have

1. $$\int_E cf\, d\mu = c \int_E f\, d\mu$$
2. $$\int_E (f + g)\, d\mu = \int_E f\, d\mu + \int_E g\, d\mu$$

Proof.

1. The first part is clear because we know that, if $s \geq 0$ is a simple function, then $I_E(cs) = cI_E(s)$ and also that $s \leq f$ if and only if $cs \leq cf$.
2. Again we know that, if s_1 and s_2 are nonnegative simple functions, then

$$I_E(s_1 + s_2) = I_E(s_1) + I_E(s_2)$$

Now, choose a monotone sequence of simple functions

$$0 \leq s_1 \leq s_2 \leq \cdots$$

with $s_n \to f$ pointwise. Similarly, choose simple functions

$$0 \leq s_1' \leq s_2' \leq \cdots$$

with $s_n' \to g$ pointwise.

Note that

$$0 \le s_1 + s_1' \le s_2 + s_2' \le \cdots$$

and $s_n + s_n' \to f + g$ pointwise. By formula 3

$$\int_E (f + g)\, d\mu = \lim_{n \to \infty} \int_E (s_n + s_n')\, d\mu$$

$$= \lim_{n \to \infty} \left(\int_E s_n\, d\mu + \int_E s_n'\, d\mu \right)$$

$$= \lim_{n \to \infty} \int_E s_n\, d\mu + \lim_{n \to \infty} \int_E s_n'\, d\mu$$

$$= \int_E f\, d\mu + \int_E g\, d\mu \qquad \square$$

Corollary 4. Let $f_1, f_2, \ldots$ be a sequence of nonnegative measurable functions. Then $\sum_{i=1}^{\infty} f_i$ is a nonnegative measurable function and

$$\int_E \left(\sum_{n=1}^{\infty} f_n \right) d\mu = \sum_{n=1}^{\infty} \int_E f_n\, d\mu \qquad \text{for } E \in \mathcal{F}$$

Proof. Let $F_n = \sum_{k=1}^{n} f_k$. Then $F_1 \le F_2 \le \cdots$, and the F_n's are all measurable. Now apply the monotone convergence theorem and theorem 3. $\qquad \square$

So far we have integrated only *nonnegative* measurable functions. When extending our definition to more general measurable functions, we must beware of the problem of adding $+\infty$ to $-\infty$.

Let f be an arbitrary measurable function from X into the extended real numbers. Recall that

$$f_+ = \max(f, 0) \qquad \text{and} \qquad f_- = \max(-f, 0)$$

are nonnegative measurable functions with

$$f = f_+ - f_-$$

Lemma 5. The following two conditions are equivalent.

1. $\displaystyle \int_E |f|\, d\mu < \infty$

2. $\displaystyle \int_E f_+\, d\mu < \infty \qquad \text{and} \qquad \int_E f_-\, d\mu < \infty$

Proof. Notice that $|f| = f_+ + f_-$, so

(4)
$$\int_E |f| \, d\mu = \int_E f_+ \, d\mu + \int_E f_- \, d\mu \qquad \triangledown$$

Definition 6. A measurable function f is *integrable* over E if either of the equivalent conditions of lemma 5 holds. In this case we write $f \in \mathscr{L}(\mu, E)$ or $f \in \mathscr{L}(\mu)$ on E. If $E = X$ we write $f \in \mathscr{L}(\mu)$. For $f \in \mathscr{L}(\mu, E)$ we define

(5)
$$\int_E f \, d\mu = \int_E f_+ \, d\mu - \int_E f_- \, d\mu$$

Theorem 7. (Linearity) Let $f, g \in \mathscr{L}(\mu, E)$ and $c \in \mathbf{R}$. Then

a. $cf \in \mathscr{L}(\mu, E)$ and $\displaystyle \int_E cf \, d\mu = c \int_E f \, d\mu$

b. $f + g \in \mathscr{L}(\mu, E)$ and $\displaystyle \int_E (f + g) \, d\mu = \int_E f \, d\mu + \int_E g \, d\mu$

Proof.

a. If $c \geq 0$, then $c(f_+) = (cf)_+$ and $(cf)_- = c(f_-)$, so

$$\int_E cf \, d\mu = \int_E cf_+ \, d\mu - \int_E cf_- \, d\mu$$

$$= c \int_E f_+ \, d\mu - c \int_E f_- \, d\mu$$

$$= c \int_E f \, d\mu$$

A similar argument treats the case of $c < 0$.

b. Let $h = f + g$. We first deal with the special case that none of f, g, or h changes sign on E. The six subcases are:
1. $f \geq 0, g \geq 0, h \geq 0$ on E
2. $f \leq 0, g \leq 0, h \leq 0$ on E
3. $f \geq 0, g \leq 0, h \geq 0$ on E
4. $f \geq 0, g \leq 0, h \leq 0$ on E
5. $f \leq 0, g \geq 0, h \geq 0$ on E
6. $f \leq 0, g \geq 0, h \leq 0$ on E

Case 1 has been dealt with previously. *Case 2* can be reduced to case 1 because we can rewrite the formula as

$$\int_E (-h)\, d\mu = \int_E (-f)\, d\mu + \int_E (-g)\, d\mu$$

Case 3: rewriting $h = f + g$ as $f = h + (-g)$ reduces this to case 1. Similarly, cases 4, 5, and 6 can be reduced to case 1.

Now to complete the proof we write $E = E_1 \cup E_2 \cup \cdots \cup E_6$ so that E_i is the set for which case i holds, $i = 1, 2, \ldots, 6$. Then, because $\int_E f\, d\mu = \sum_{i=1}^6 \int_{E_i} f\, d\mu$, and similarly for g and h, the theorem follows by applying it to each E_i separately and summing. $\qquad \square$

Corollary 8. (Monotonicity) Let $f, g \in \mathcal{L}(\mu, E)$ with $f \le g$. Then

(6)
$$\int_E f\, d\mu \le \int_E g\, d\mu$$

Proof. Because $f \le g$, $g - f \ge 0$, so

$$0 \le \int_E (g - f)\, d\mu = \int_E g\, d\mu - \int_E f\, d\mu$$

by linearity. $\qquad \square$

Corollary 9. If $f \in \mathcal{L}(\mu, E)$, then

(7)
$$\left| \int_E f\, d\mu \right| \le \int_E |f|\, d\mu$$

Proof. Because $f \le |f|$, by monotonicity

$$\int_E f\, d\mu \le \int_E |f|\, d\mu$$

Similarly, $-f \le |f|$, so

$$-\int_E f\, d\mu \le \int_E |f|\, d\mu$$

That is,

$$\left| \int_E f\, d\mu \right| \le \int_E |f|\, d\mu \qquad \square$$

We end this section with the other convergence theorems mentioned above.

Lemma 10. (Fatou's lemma) Let $f_1, f_2, \ldots$ be a sequence of nonnegative measurable functions, and let $f = \liminf f_n$. Then

$$(8) \qquad \int_E f \, d\mu \le \liminf \int_E f_i \, d\mu$$

Proof. Let $g_k = \inf_{n \ge k} f_n$ and $a_k = \inf_{n \ge k} \int_E f_k \, d\mu$. Then

$$g_1 \le g_2 \le \cdots \qquad \text{and} \qquad a_1 \le a_2 \le \cdots$$

By definition

$$f = \lim_{k \to \infty} g_k \qquad \text{and} \qquad \liminf \int_E f_n \, d\mu = \lim_{k \to \infty} a_k$$

Notice that $a_k \ge \int_E g_k \, d\mu$ because $g_k \le f_n$ for $n \ge k$. Hence, by the monotone convergence theorem,

$$\int_E f \, d\mu = \lim_{k \to \infty} \int_E g_k \, d\mu \le \lim_{k \to \infty} a_k = \liminf \int_E f_k \, d\mu \qquad \square$$

Theorem 11. (Lebesgue dominated convergence theorem) Let $f_1, f_2, f_3, \ldots$ be a sequence of measurable functions, and let $E \in \mathscr{F}$. Assumptions:

1. $\lim_{n \to \infty} f_n(x)$ exists for all $x \in E$.
2. There is a nonnegative measurable function $g \in \mathscr{L}(\mu, E)$ with $g \ge |f_n|$ on E, $n = 1, 2, \ldots$.

Conclusion: The function $f(x) = \lim_{n \to \infty} f_n(x)$ is integrable and

$$\int_E \lim_{n \to \infty} f_n \, d\mu = \lim_{n \to \infty} \int_E f_n \, d\mu$$

Proof. By Fatou's lemma,

$$\int_E |f| \, d\mu = \int_E \liminf |f_n| \, d\mu$$

$$\le \liminf \int_E |f_n| \, d\mu \le \int g \, d\mu$$

Hence $f \in \mathscr{L}(\mu, E)$.

Now we note that $g + f_n$ is nonnegative, so by Fatou's lemma

$$\int_E \liminf(g + f_n) \, d\mu \le \liminf \int_E (g + f_n) \, d\mu$$

But
$$\liminf(g + f_n) = g + \lim_{n \to \infty} f_n = g + f$$

and
$$\liminf \int_E (g + f_n)\, d\mu = \int_E g\, d\mu + \liminf \int_E f_n\, d\mu$$

so we have
$$\int_E f\, d\mu \le \liminf \int_E f_n\, d\mu$$

Repeating this argument, with $g + f_n$ replaced by $g - f_n$, we get

$$-\int_E f\, d\mu \le \liminf \left(-\int_E f_n\, d\mu \right)$$

$$= -\limsup \left(\int_E f_n\, d\mu \right)$$

so
$$\int_E f\, d\mu \ge \limsup \left(\int_E f_n\, d\mu \right)$$

Combining these results

$$\limsup \left(\int_E f_n\, d\mu \right) \le \int_E f\, d\mu \le \liminf \left(\int_E f_n\, d\mu \right)$$

But it is always true that $\liminf \le \limsup$, thus we get equality:

$$\int_E f\, d\mu = \liminf \left(\int_E f_n\, d\mu \right) = \limsup \left(\int_E f_n\, d\mu \right) = \lim \left(\int_E f_n\, d\mu \right) \qquad \square$$

Corollary 12.   Let $f_1, f_2, \ldots$ be a sequence in $\mathscr{L}(\mu, E)$ with

$$\sum_{n=1}^{\infty} \int_E |f_n|\, d\mu < \infty$$

Then

1. $\sum_{n=1}^{\infty} f_n$ converges absolutely a.e. on E and is integrable on E.
2. $\int_E \sum_{n=1}^{\infty} f_n\, d\mu = \sum_{n=1}^{\infty} \int_E f_n\, d\mu$.

Proof.   Let $g = \sum_{n=1}^{\infty} |f_n|$. Corollary 4 tells us that

$$\int_E g\, d\mu = \sum_{n=1}^{\infty} \int_E |f_n|\, d\mu < \infty$$

so $g \in \mathscr{L}(\mu, E)$. In particular, g is finite a.e. on E, so $\sum_{n=1}^{\infty} f_n$ converges absolutely a.e. on E. To prove part 2 let $F_n = \sum_{k=1}^{n} f_k$. Then $|F_n| \le \sum_{k=1}^{n} |f_k| \le$

g so, by the dominated convergence theorem, $\sum_{n=1}^{\infty} f_n = \lim_{n\to\infty} F_n$ is in $\mathscr{L}(\mu, E)$ and

$$\int_E \left(\sum_{n=1}^{\infty} f_n\right) d\mu = \int_E \lim_{n\to\infty} F_n \, d\mu = \lim_{n\to\infty} \int_E F_n \, d\mu = \sum_{n=1}^{\infty} \int_E f_n \, d\mu \qquad \square$$

Exercises for §2.3

1. Let $(X, \mathscr{F}, \mu)$ be a measure space and f be a bounded nonnegative measurable function. Show that

$$\int_X f \, d\mu = \lim_{n\to\infty} \sum_{k=1}^{n2^n} \left(\frac{k-1}{2^n}\right) \mu\left(\left\{x \in X; \frac{k-1}{2^n} \le f(x) < \frac{k}{2^n}\right\}\right)$$

   This formula is Lebesgue's original definition of the Lebesgue integral. (*Hint:* See formula 3, page 74.)

2. Let $f_n : \mathbf{R} \to \mathbf{R}$ be $1/n$ times the characteristic function of the interval $(0, n)$. Show that $f_n \to 0$ uniformly but that $\int f_n \, d\mu_L = 1$. Why isn't this example a counterexample to the Lebesgue dominated convergence theorem?

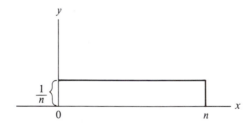

3. Compare $\int \liminf f_n \, d\mu_L$ and $\liminf \int f_n \, d\mu_L$ for the sequence f_n in exercise 2. Can the inequality in Fatou's lemma be replaced by an equality?

4. For $n = 1, 3, 5, \ldots$ let f_n be the characteristic function of the interval $(0, \frac{1}{2})$, and for $n = 2, 4, 6, \ldots$ let f_n be the characteristic function of the interval $(\frac{1}{2}, 1)$. Compare $\int \liminf f_n \, d\mu_L$ and $\liminf \int f_n \, d\mu_L$.

5. Let $(X, \mathscr{F}, \mu)$ be a measure space. A measurable function $f : X \to \mathbf{R}$ is mean-square integrable if $\int f^2 \, d\mu < \infty$. Show that, if $\mu(X) < \infty$, every mean-square integrable function is integrable. (*Hint:* Consider separately the integral of $|f|$ over the set where $|f| \ge 1$ and the integral over the set where $|f| < 1$.)

6. Let $X = I$, $\mathscr{F}$ the Borel sets, and μ Lebesgue measure. Show that there exists an integrable function on X that is not mean-square integrable.

7. Let $X = \mathbf{R}$, $\mathscr{F}$ the Borel sets, and μ Lebesgue measure. Show that there exists a mean-square integrable function on X that is not integrable.

8. Let $(X, \mathscr{F}, \mu)$ be a measure space and f_n, $n = 1, 2, \ldots$, a sequence of measurable functions. Then f_n is said to converge to zero in measure if, for all $\varepsilon > 0$,

$$\mu(\{x \in X; |f_n(x)| > \varepsilon\}) \to 0$$

as $n \to \infty$. (Compare with exercise 11 of §2.1.) Show that, if $\int |f_n| \, d\mu \to 0$, then f_n converges to zero in measure. Show that the converse is not true. (*Hint:* See exercise 2.)

9. Let $f : \mathbf{R} \to \mathbf{R}$ be an integrable function. Show that, if

$$\int_I f \, d\mu = 0$$

for every subinterval I of the real line, then $f = 0$ a.e.

10. Let J be a finite subinterval of the real line and $f : J \to \mathbf{R}$ an integrable function. Show that for every $\varepsilon > 0$ there exists a step function g such that

$$\int_J |f - g| \, d\mu < \varepsilon$$

(*Hint:* See §2.2, exercise 7.)

11. Let $(X, \mathscr{F}, \mu)$ be a measure space and f and g measurable functions. Show that, if f is integrable and g is bounded and measurable, then fg is integrable.

12. Let $(X, \mathscr{F}, \mu)$ be a measure space and f a nonnegative measurable function. For $A \in \mathscr{F}$ let

$$\mu_f(A) = \int_A f \, d\mu$$

Show that μ_f is a measure on $\mathscr{F}$. Moreover, show that, if g is a nonnegative measurable function, then for $E \in \mathscr{F}$

$$\int_E g \, d\mu_f = \int_E gf \, d\mu$$

13. **a.** In exercise 12, let $A \in \mathscr{F}$ and let f be the characteristic function of the set A. Describe the measure μ_f.
    b. Suppose, more generally, that f is a simple function; that is

(*)
$$f = \sum_{i=1}^k c_i 1_{A_i}$$

Describe the measure μ_f.

14. **a.** In exercise 12, show that, if f is bounded, $\mathscr{L}(E, \mu_f)$ contains $\mathscr{L}(E, \mu)$. Moreover, show that, for $g \in \mathscr{L}(E, \mu)$

$$\int_E g \, d\mu_f = \int_E gf \, d\mu$$

b. Show that $\mathscr{L}(E, \mu_f)$ need not necessarily contain $\mathscr{L}(E, \mu)$ if f is not bounded.

15. a. Let $(X, \mathscr{F}, \mu)$ be a measure space and $f : X \to \mathbf{R}$ a measurable function. For every Borel set $A \subseteq \mathbf{R}$, let

$$v_f(A) = \mu(f^{-1}(A))$$

Show that this formula defines a measure v_f on the Borel sets of $\mathbf{R}$. Moreover, show that, if μ is a probability measure, so is v_f.

b. If f is the function (∗), describe the measure v_f.

16. A function $g : \mathbf{R} \to \mathbf{R}$ is *Borel measurable* if, for every Borel set $A \subseteq \mathbf{R}$, $g^{-1}(A)$ is a Borel set. Let v_f be the measure in exercise 15 and let g be a nonnegative Borel-measurable function on the real line. Show that

(∗∗) $$\int_{\mathbf{R}} g \, d\mu_f = \int_X g(f(x)) \, d\mu$$

(*Hint:* What does equation (∗∗) say when g is a simple function?)

§2.4 Lebesgue Integration versus Riemann Integration

By the results of §2.3 we can now integrate complicated limits and sums of series. What about simple integrals? We will show that in the Lebesgue theory, just as in the Riemann theory, these integrals can be evaluated by the second fundamental theorem of calculus; that is, for the Riemann integral one has the following theorem.

Theorem. Let g be a continuous function on an interval $[a, b] \subset \mathbf{R}$. Then g has an antiderivative G and

$$\int_a^b g \, dx = G(b) - G(a)$$

where $\int_a^b g \, dx$ denotes the Riemann integral of g on the interval $[a, b]$.

We will show that the same is true for the Lebesgue integral when our measure space $(X, \mathscr{F}, \mu)$ is $X = [a, b]$, $\mu = \mu_L$, and $\mathscr{F}$ is the field of Lebesgue-measurable subsets of X. Rather than prove the fundamental theorem of calculus directly for the Lebesgue integral, we prove a much more general theorem, as follows.

Theorem 1.   Let f be a bounded Riemann-integrable function on $[a,b]$ with Riemann integral $\int_a^b f\,dx$. Then $f \in \mathcal{L}(\mu_L, [a,b])$ and

(1)
$$\int_a^b f\,dx = \int_{[a,b]} f\,d\mu_L$$

Before proving this theorem, let us recall how the Riemann integral is defined.

Riemann Integral

A *partition* P of $[a,b]$ is a finite, ordered sequence of points

$$a = x_0 < x_1 < x_2 < \cdots < x_N = b$$

The maximum of the numbers $x_{i+1} - x_i, i = 0, \ldots, N-1$ is denoted $m(P)$ and is called the *mesh width* of the partition.

Fix a partition P of $[a,b]$, and let f be a bounded function on $[a,b]$. Let

$$M_i = \sup\{f(x); x_{i-1} \le x \le x_i\}$$

and

$$m_i = \inf\{f(x); x_{i-1} \le x \le x_i\}$$

Define

$$U(f,P) = \sum_{i=1}^N M_i(x_i - x_{i-1})$$

and

$$L(f,P) = \sum_{i=1}^N m_i(x_i - x_{i-1})$$

to be the *upper and lower Riemann sums*, respectively.

Notice that $L(f,P) \le U(f,P)$. In fact, it is true that, if P_1 and P_2 are any two partitions of $[a,b]$, then $L(f,P_1) \le U(f,P_2)$. To see this we introduce the notion of *refinement*. A partition P' is called a *refinement* of P if the ordered sequence of points comprising P' contains the points of P as well as some additional points; that is, $P' = P$ plus additional points. Clearly, if P' is a refinement of P, then $L(f,P') \ge L(f,P)$ and $U(f,P') \le U(f,P)$. Now, if P_1 and P_2 are two partitions of $[a,b]$, let P be a partition of $[a,b]$ that refines both P_1 and P_2 simultaneously. Then

(2) $L(f,P_1) \le L(f,P) \le U(f,P) \le U(f,P_2)$

We define

$$(3) \quad \int_a^{\overline{b}} f \, dx = \inf\{U(f, P); P \text{ is a partition of } [a, b]\}$$

$$\int_{\underline{a}}^b f \, dx = \sup\{L(f, P); P \text{ is a partition of } [a, b]\}$$

Note that by inequality 2

$$(4) \quad \int_{\underline{a}}^b f \, dx \leq \int_a^{\overline{b}} f \, dx$$

Definition 2.   The function f is *Riemann integrable* if $\int_a^{\overline{b}} f \, dx = \int_{\underline{a}}^b f \, dx$. In this case we define

$$(5) \quad \int_a^b f \, dx = \int_{\underline{a}}^b f \, dx = \int_a^{\overline{b}} f \, dx$$

To compare the Riemann integral with the Lebesgue integral, we will first show the following lemma.

Lemma 3.   There exists a sequence of partitions $P_1, P_2, P_3, \ldots$ such that

1. P_k is a refinement of P_{k-1}.
2. The monotone sequence

$$U(f, P_1) \geq U(f, P_2) \geq \cdots$$

   converges to $\int_a^{\overline{b}} f \, dx$.
3. The monotone sequence

$$L(f, P_1) \leq L(f, P_2) \leq \cdots$$

   converges to $\int_{\underline{a}}^b f \, dx$.
4. The mesh width $m(P_k) \to 0$ as $k \to \infty$.

Proof.   We describe how to define P_k. Let P_k' be a partition with

$$U(f, P_k') \leq \int_a^{\overline{b}} f \, dx + \frac{1}{k}$$

and P_k'' a partition with

$$L(f, P_k'') \geq \int_a^b f \, dx - \frac{1}{k}$$

We know that P_k' and P_k'' exist by the definition of $\overline{\int_a^b} f \, dx$ and $\underline{\int_a^b} f \, dx$. Let P_k be a partition with $m(P_k) < 1/k$ refining P_k', P_k'', and P_{k-1} all at once. Then

$$L(f, P_k) \geq L(f, P_k'') \geq \int_a^b f \, dx - \frac{1}{k}$$

and

$$U(f, P_k) \leq U(f, P_k') \leq \overline{\int_a^b} f \, dx + \frac{1}{k}$$

So $\qquad \lim_{k \to \infty} L(f, P_k) = \underline{\int_a^b} f \, dx \qquad$ and $\qquad \lim_{k \to \infty} U(f, P_k) = \overline{\int_a^b} f \, dx \qquad \triangledown$

For a fixed partition P, define the simple functions

$$L_P(x) = \begin{cases} f(a) & \text{at } x = a \\ m_1 & \text{on } x_0 < x \leq x_1 \\ m_2 & \text{on } x_1 < x \leq x_2 \\ \vdots & \vdots \\ m_N & \text{on } x_{N-1} < x \leq x_N \end{cases}$$

and

$$U_P(x) = \begin{cases} f(a) & \text{at } x = a \\ M_1 & \text{on } x_0 < x \leq x_1 \\ M_2 & \text{on } x_1 < x \leq x_2 \\ \vdots & \vdots \\ M_N & \text{on } x_{N-1} < x \leq x_N \end{cases}$$

Notice that $L_p(x) \leq f(x) \leq U_P(x)$ and

$$\int_{[a,b]} L_P(x) \, d\mu_L = \sum_{i=1}^N m_i(x_i - x_{i-1}) = L(f, P)$$

$$\int_{[a,b]} U_P(x) \, d\mu_L = \sum_{i=1}^N M_i(x_i - x_{i-1}) = U(f, P)$$

See figure on page 86.

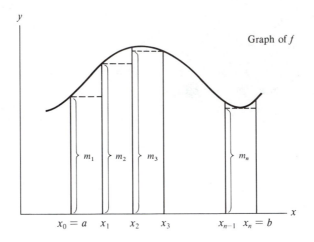

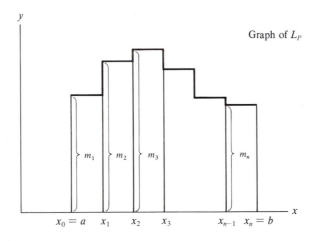

Also, if P' is a refinement of P, then

$$U_{P'}(x) \leq U_P(x) \qquad \text{and} \qquad L_{P'}(x) \geq L_P(x)$$

Now choose a sequence of P_k's as in lemma 3 and let

$$L_k = L_{P_k} \qquad \text{and} \qquad U_k = U_{P_k}$$

Then

$$L_1 \leq L_2 \leq \cdots \leq f \leq \cdots \leq U_2 \leq U_1$$

Let $L(x) = \lim_{n \to \infty} L_n(x)$ and $U(x) = \lim_{n \to \infty} U_n(x)$. By construction, $L(x) \leq f(x) \leq U(x)$, and, by the monotone convergence theorem,

$$\lim_{n \to \infty} \int_{[a,b]} L_n \, d\mu_L = \int_{[a,b]} L \, d\mu_L$$

and

$$\lim_{n \to \infty} \int_{[a,b]} U_n \, d\mu_L = \int_{[a,b]} U \, d\mu_L$$

Now we know that

$$\int_{[a,b]} L_n \, d\mu_L = L(f, P_n) \qquad \text{and} \qquad \int_{[a,b]} U_n \, d\mu_L = U(f, P_n)$$

and we chose the P_n's so that

$$\lim_{n \to \infty} L(f, P_n) = \underline{\int_a^b} f \, dx \qquad \text{and} \qquad \lim_{n \to \infty} U(f, P_n) = \overline{\int_a^b} f \, dx$$

Hence, we conclude

$$\int_{[a,b]} L \, d\mu_L = \underline{\int_a^b} f \, dx \qquad \text{and} \qquad \int_{[a,b]} U \, d\mu_L = \overline{\int_a^b} f \, dx$$

To prove the theorem, we assume that f is Riemann integrable; that is,

$$\underline{\int_a^b} f \, dx = \overline{\int_a^b} f \, dx$$

In other words,

$$\int_{[a,b]} L \, d\mu_L = \int_{[a,b]} U \, d\mu_L = \int_a^b f \, dx$$

Thus $\int_{[a,b]} (U - L) \, d\mu_L = 0$. But $U \geq L$ so $U - L \geq 0$; hence, $U - L = 0$ a.e.

Now $\qquad\qquad L \leq f \leq U \qquad$ so $\qquad f = U = L \quad$ a.e.

and f is Lebesgue integrable with

$$\int_{[a,b]} f \, d\mu_L = \int_a^b f \, dx \qquad\qquad \triangledown$$

Remark. The standard notation for the Riemann integral of a function f over an interval $[a, b]$ is the notation we have been using—namely,

(6)
$$\int_a^b f(x) \, dx$$

Unfortunately, the Lebesgue integral has no such standard notation. Heretofore we have been using the notation

(7)
$$\int_{[a,b]} f \, d\mu_L$$

or

(8)
$$\int_I f \, d\mu_L$$

with I denoting the interval $[a, b]$.

Now that we've shown that the Lebesgue integral and the Riemann integral are identical for Riemann-integrable functions, we will be less methodical with our notation. We will sometimes use display 6 for Lebesgue integrals and will sometimes use displays 7 and 8 with the subscript deleted from μ_L when it is clear from the context that μ is μ_L.

Exercises for §2.4

1. Compute the Lebesgue integral

$$\int_I f \, d\mu_L$$

   of the following functions: x^2, x^3, $\sin x$, e^x, $x \log x$. (You are encouraged to use the tools of elementary calculus in making these computations.)

2. In the proof of theorem 1, choose an $x \in [a, b]$ that is not equal to any of the partition points of any of the P_k's. Show that f is continuous at x if and only if $U(x) = L(x)$.

3. Conclude from exercise 2 that, if a bounded function f on the interval $[a, b]$ is Riemann integrable, then it is continuous almost everywhere.

4. **a.** Conversely, suppose that f is a bounded function that is continuous almost everywhere on the interval $[a, b]$. Conclude from exercise 2 that

   $$\lim U(f, P_k) = \lim L(f, P_k)$$

   b. Deduce that, if f is bounded and continuous a.e. on $[a, b]$, then it is Riemann integrable.

5. (Improper integrals)
   a. Let f be a nonnegative measurable function on the interval $J = (0, 1]$. Suppose that f is Riemann integrable on all of the intervals $[a, 1]$, $0 < a < 1$. Show that

   $$\int_J f \, d\mu_L = \lim_{a \to 0} \int_a^1 f(x) \, dx$$

   with the integral on the right being the Riemann integral.
   b. Compute the Lebesgue integral over J of the function $f(x) = 1/\sqrt{x}$.

6. a. Let f be a nonnegative measurable function on the interval $J = [1, \infty)$. Suppose that f is Riemann integrable on all of the intervals $[1, a]$, $a > 1$. Show

$$\int_J f \, d\mu_L = \lim_{a \to \infty} \int_1^a f(x) \, dx$$

b. Compute the Lebesgue integral over J of the function $f(x) = 1/x^2$.

7. Let $f(x) = (\sin x)/x$. Show that

$$\lim_{a \to \infty} \int_1^a f(x) \, dx$$

exists. On the other hand, show that $f(x)$ is *not* Lebesgue integrable over the interval $[1, \infty)$.

8. Let $f : \mathbf{R} \to \mathbf{R}$ be continuous and have a continuous first derivative f' that is positive everywhere. If μ is Lebesgue measure on $\mathbf{R}$, show that the measure v_f of §2.3, exercise 15, is of the form

$$v_f(A) = \int_A g \, d\mu \qquad \text{where } 1/g(x) = f'(f^{-1}(x))$$

§2.5 Fubini Theorem

Section 2.4 and the convergence theorems of §2.3 allow us to compute a number of integrals on subsets of $\mathbf{R}$. In order to compute integrals on subsets of $\mathbf{R}^n$, we must justify the use of iterated integrals. This is the purpose of Fubini's theorem.

The general situation is as follows: Let $(X, \mathcal{M}, \mu)$ and $(Y, \mathcal{N}, v)$ be two measure spaces. Let $X \times Y$ denote the space

(1) $$X \times Y = \{(x, y); x \in X, y \in Y\}$$

If $A \subset X$ and $B \subset Y$, then $A \times B \subset X \times Y$. On the other hand, you should notice that most subsets of $X \times Y$ are not of this form.

Definition 1. $A \times B \subset X \times Y$ is called a *product set* if $A \in \mathcal{M}$ and $B \in \mathcal{N}$. The smallest σ-field in $X \times Y$ containing all product sets $A \times B$ is denoted $\mathcal{M} \times \mathcal{N}$.

In a moment we will define a measure on $\mathcal{M} \times \mathcal{N}$. First we must describe sets in $\mathcal{M} \times \mathcal{N}$ in terms of $\mathcal{M}$ and $\mathcal{N}$.

Definition 2. For $E \subset X \times Y$ and $x \in X$ fixed, let $E_x = \{y \in Y; (x, y) \in E\}$. Then E_x is called the *x-slice* of E.

Notice that, if E and F are subsets of $X \times Y$, then

$$(E \cap F)_x = E_x \cap F_x$$

(2)
$$(E - F)_x = E_x - F_x$$

$$E_x^c = (E^c)_x$$

and, if $E_1, E_2, E_3, \ldots$ are subsets of $X \times Y$, then

$$\left(\bigcup_{n=1}^{\infty} E_n \right)_x = \bigcup_{n=1}^{\infty} (E_n)_x$$

Proposition 3.   If $E \in \mathcal{M} \times \mathcal{N}$, then $E_x \in \mathcal{N}$.

Proof.   Fix $x \in X$ and let $\mathcal{S}_x$ be the collection of all sets $E \subseteq X \times Y$ with $E_x \subseteq \mathcal{N}$. Note that

1. $\mathcal{S}_x$ contains all product sets $A \times B$.
2. $\mathcal{S}_x$ is a σ-field.

Thus $\mathcal{S}_x$ contains the smallest σ-field containing all product sets—namely, $\mathcal{M} \times \mathcal{N}$. $\square$

Corollary 4.   Let $f : X \times Y \to \mathbf{R} \cup \{\pm\infty\}$ be measurable with respect to $\mathcal{M} \times \mathcal{N}$. For $x_0 \in X$ fixed, define $f_{x_0} : Y \to \mathbf{R}$ by $f_{x_0}(y) = f(x_0, y)$. Then, for each $x_0 \in X$, f_{x_0} is a measurable function on Y.

Proof.   Fix $x_0 \in X$. If $a \in \mathbf{R}$ we need to show that

$$\{ y \in Y; \ f_{x_0}(y) < a \} \in \mathcal{N}$$

Let $E = \{(x, y) \in X \times Y; \ f(x, y) < a\}$. $E \in \mathcal{M} \times \mathcal{N}$ because f is measurable and $\{ y \in Y; \ f_{x_0}(y) < a \} = E_{x_0} \in \mathcal{N}$ by proposition 3. $\square$

Remark.   We could just as easily have studied y-slices as x-slices. The corresponding proposition and corollary are obviously true for y-slices as well.

Thus far we have made no assumptions about the measure spaces $(X, \mathcal{M}, \mu)$ and $(Y, \mathcal{N}, \nu)$. We will now assume that both of these spaces are σ-finite. (Recall that a measure space $(X, \mathcal{M}, \mu)$ is σ-finite if there exist $X_i \subset \mathcal{M}$, $i = 1, 2, 3, \ldots$ with $\mu(X_i) < \infty$ and

$$\bigcup_{i=1}^{\infty} X_i = X$$

(See definition 32 in §1.3.)

We now use the measures μ and ν to define a measure on $\mathcal{M} \times \mathcal{N}$. First let's recall how one computes the areas of regions in the plane in elementary calculus. Consider the region in the figure below.

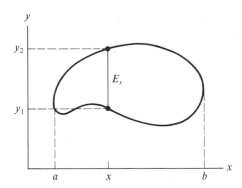

For each point x on the interval $[a, b]$, we let $\phi_E(x)$ be the length of the interval E_x; that is, $\phi_E(x) = y_2 - y_1$. For reasonable-looking regions, such as the one we've drawn here, $\phi_E(x)$ is a continuous function of x, so the Riemann integral

$$(3) \qquad \int_a^b \phi_E(x)\, dx$$

is well defined. In elementary calculus one proves that display 3 gives the area of the region E (or uses the integral as the definition of area). To define a measure on the product σ-field, $\mathcal{M} \times \mathcal{N}$, we will mimic this process. Let $E \in \mathcal{M} \times \mathcal{N}$. For each $x \in X$, $E_x \in \mathcal{N}$, so we can define a function $\phi_E : X \to \mathbf{R}$ by

$$(4) \qquad \phi_E(x) = \nu(E_x)$$

We would like to define the measure of E to be the integral

$$(5) \qquad \int_X \phi_E(x)\, d\mu$$

To use this, we have to check that $\phi_E(x)$ is a measurable function on X. Proof of this fact requires a general argument about σ-fields called the $\pi-\lambda$ theorem.

Definition 5. Let Z be a set and let $\mathscr{S}$ be a collection of subsets of Z. $\mathscr{S}$ is called a *λ-system* if the following three properties hold.

λ1. $Z \in \mathscr{S}$.

λ2. If $E_1 \subset E_2 \subset E_3 \subset \cdots$ is an increasing sequence with each $E_n \in \mathscr{S}$, then

$$\bigcup_{n=1}^{\infty} E_n \in \mathscr{S}$$

λ3. If $E, F \in \mathscr{S}$ and $E \subset F$, then $F - E \in \mathscr{S}$.

Definition 6.   A collection $\mathscr{A}$ of subsets of Z is called a *π-system* if

π**1.** $A, B \in \mathscr{A} \Rightarrow A \cap B \in \mathscr{A}$.

Theorem 7.   (π–λ theorem) If $\mathscr{S}$ is a λ-system and $\mathscr{A}$ is a π-system with $\mathscr{A} \subseteq \mathscr{S}$, then the smallest σ-field containing $\mathscr{A}$, $\sigma(\mathscr{A})$, is contained in $\mathscr{S}$.

Proof.*   First notice that, if a collection l of subsets of Z is both a λ-system and a π-system, then it must be a σ-field.

Now let $l(\mathscr{A})$ be the smallest λ-system containing $\mathscr{A}$. If we can show that $l(\mathscr{A})$ is also a π-system, then we will know it is a σ-field. Because $\sigma(\mathscr{A})$ is the smallest σ-field containing $\mathscr{A}$ and $l(\mathscr{A})$ is the smallest λ-system containing $\mathscr{A}$, we will have

$$\sigma(\mathscr{A}) \subseteq l(\mathscr{A}) \subseteq \mathscr{S}$$

which proves the theorem.

To see that $l(\mathscr{A})$ is a π-system, we need to show that if $A, B \in l(\mathscr{A})$ then $A \cap B \in l(\mathscr{A})$.

For $A \subset Z$ let

$$\mathscr{G}_A = \{B \subset Z;\ A \cap B \in l(\mathscr{A})\}$$

Notice that if $A \in l(\mathscr{A})$, then $\mathscr{G}_A$ is a λ-system. Indeed, the three properties are satisfied as follows.

λ**1.** If $A \in l(\mathscr{A})$, then $Z \in \mathscr{G}_A$, because $A \cap Z = A$.
λ**2.** If $A \in l(\mathscr{A})$ and $E_1 \subset E_2 \subset E_3 \subset \cdots$ are in $\mathscr{G}_A$, then $(A \cap E_1) \subset (A \cap E_2) \subset \cdots$ is an increasing sequence in $l(\mathscr{A})$, so $\bigcup_{n=1}^{\infty}(A \cap E_n) = A \cap (\bigcup_{n=1}^{\infty} E_n)$ is also in $l(\mathscr{A})$; that is, $\bigcup_{n=1}^{\infty} E_n \in \mathscr{G}_A$.
λ**3.** If $A \in l(\mathscr{A})$ and $E,\ F \in \mathscr{G}_A$ with $E \subset F$, then $A \cap E,\ A \cap F \in l(\mathscr{A})$, and $(A \cap E) \subset (A \cap F)$. So $A \cap (F - E) = (A \cap F) - (A \cap E) \in l(\mathscr{A})$; that is, $F - E \in \mathscr{G}_A$.

Furthermore, if $A,\ B \in \mathscr{A}$, then $A \cap B \in \mathscr{A} \subset l(\mathscr{A})$, so $B \in \mathscr{G}_A$; that is, $\mathscr{A} \subset \mathscr{G}_A$ when $A \in \mathscr{A}$. Thus we see that $l(\mathscr{A}) \subset \mathscr{G}_A$ when $A \in \mathscr{A}$, by the minimality of $l(\mathscr{A})$. In other words, we now know that, if $A \in \mathscr{A}$ and $B \in l(\mathscr{A})$, then $A \cap B \in l(\mathscr{A})$.

Thus, if $B \in l(\mathscr{A})$ we have shown that $\mathscr{A} \subset \mathscr{G}_B$. Again using the minimality of $l(\mathscr{A})$, we get $l(\mathscr{A}) \subset \mathscr{G}_B$ for $B \in l(\mathscr{A})$; that is, $A \cap B \in l(\mathscr{A})$ when A and B are elements of $l(\mathscr{A})$. This is property π1 for $l(\mathscr{A})$. $\square$

With the π–λ theorem we can now prove the following proposition.

---

*This proof is rather technical, and you might want to skip it when reading this material for the first time.

Proposition 8. If $E \in \mathcal{M} \times \mathcal{N}$ and $\phi_E : X \to \mathbf{R}$ is defined by $\phi_E(x) = v(E_x)$, then ϕ_E is measurable.

Proof. First assume that $v(Y) < \infty$. By the π–λ theorem, the proposition will follow from the following three facts.

a. Let $\mathcal{S}$ be the collection of sets E such that ϕ_E is measurable. Then $\mathcal{S}$ is a λ-system.
b. All product sets, $A \times B$, with $A \in \mathcal{M}$ and $B \in \mathcal{N}$, are in $\mathcal{S}$.
c. The collection $\mathcal{A}$ of product sets is a π-system.

We prove these statements as follows.

a. λ1. $\phi_{X \times Y}(x) = v(Y)$ for all x; hence, is clearly measurable.
   λ2. Let $E_1 \subseteq E_2 \subseteq \cdots$ be an increasing sequence with $E_n \in \mathcal{S}$; that is, $\phi_{E_n}(x)$ is measurable. Let $E = \bigcup_{n=1}^{\infty} E_n$. We need to show that $\phi_E(x)$ is measurable.
   Now

$$\phi_E(x) = v(E_x) = v\left[\bigcup_{n=1}^{\infty} (E_n)_x\right] = \lim_{n \to \infty} v[(E_n)_x] = \lim_{n \to \infty} \phi_{E_n}(x)$$

   Thus ϕ_E is the pointwise limit of measurable functions and so is measurable as well.
   λ3. Let $E, F \in \mathcal{S}$ with $E \subset F$. Then $F = E \cup (F - E)$ is a disjoint union, and $F_x = E_x \cup (F - E)_x$ is also disjoint. Thus $v(F_x) = v(E_x) + v[(F - E)_x]$; that is

$$\phi_{F-E}(x) = \phi_F(x) - \phi_E(x)$$

   Because ϕ_E and ϕ_F are measurable, ϕ_{F-E} is too.
b. Let $E = A \times B$ be a product set. Then

$$E_x = \begin{cases} B & \text{if } x \in A \\ \phi & \text{if } x \notin A \end{cases}$$

so

(6)
$$\phi_E(x) = \begin{cases} v(B) & \text{if } x \in A \\ 0 & \text{if } x \notin A \end{cases}$$

That is, $\phi_E(x) = v(B)1_A(x)$, which we know to be measurable.
c. Let $A \times B$ and $A' \times B'$ be two product sets. Then

$$(A \times B) \cap (A' \times B') = (A \cap A') \times (B \cap B')$$

so the collection of product sets is a π-system.

Now, what happens if we drop the assumption that $v(Y) < \infty$? Because Y is σ-finite we can find a sequence of subsets $Y_1, Y_2, Y_3, \ldots$ of Y with $Y_i \in \mathcal{N}$, $v(Y_i) < \infty$, and

$$Y = \bigcup_{i=1}^{\infty} Y_i \qquad \text{(disjoint union)}$$

Let $E \in \mathcal{M} \times \mathcal{N}$. Then

$$E = \bigcup_{i=1}^{\infty} E_i \qquad \text{(disjoint union)}$$

where $E_i = E \cap (X \times Y_i)$ and

(7)
$$\phi_E(x) = \sum_{i=1}^{\infty} \phi_{E_i}(x)$$

We proved above that ϕ_{E_i} is measurable, so by equation 7, ϕ_E is measurable.

$\square$

We can now use display 5 to define a measure on $\mathcal{M} \times \mathcal{N}$.

Definition 9. Let $E \in \mathcal{M} \times \mathcal{N}$ and define

(8)
$$\pi'(E) = \int_X \phi_E(x)\, d\mu$$

to be the *product measure* of E.

Proposition 10. π' is a measure.

Proof. We must check that π' is countably additive. Let $E_1, E_2, \ldots$ be a pairwise disjoint family of sets in $\mathcal{M} \times \mathcal{N}$. Let $E = \bigcup_{n=1}^{\infty} E_n$. We need to show that

$$\pi'(E) = \sum_{n=1}^{\infty} \pi'(E_n)$$

Now, $E_x = \bigcup_{n=1}^{\infty} (E_n)_x$ is a disjoint union, so

$$v(E_x) = \sum_{n=1}^{\infty} v[(E_n)_x]$$

that is,
$$\phi_E(x) = \sum_{n=1}^{\infty} \phi_{E_n}(x)$$

We can now apply the monotone convergence theorem to get

$$\int_X \phi_E \, d\mu = \sum_{n=1}^{\infty} \int_X \phi_{E_n} \, d\mu$$

that is,
$$\pi'(E) = \sum_{n=1}^{\infty} \pi'(E_n) \qquad \square$$

Notice that we could repeat this whole procedure using y-slices instead of x-slices. Ostensibly this method gives a different measure π'' on $\mathcal{M} \times \mathcal{N}$. Our first version of Fubini's theorem is that these two methods yield the same measure.

Theorem 11. (Fubini, version 1)

$$\pi' = \pi''$$

Proof. First assume that $\mu(X)$ and $v(Y)$ are finite. By the $\pi{-}\lambda$ theorem, it is enough to establish the following three assertions.

a. Let $\mathscr{S} = \{E \in \mathcal{M} \times \mathcal{N} \, ; \, \pi'(E) = \pi''(E)\}$. Then $\mathscr{S}$ is a λ-system.
b. The product sets are in $\mathscr{S}$.
c. The product sets form a π-system.

We proved assertion c above. Let us look at assertions a and b.

a. λ1. $\pi'(X \times Y) = \mu(X)v(Y) = \pi''(X \times Y)$, so $X \times Y \in \mathscr{S}$.
λ2. Let $E_1 \subseteq E_2 \subseteq \cdots$ be an increasing sequence with $E_n \in \mathscr{S}$. We wish to show that $E = \bigcup_{n=1}^{\infty} E_n \in \mathscr{S}$. It is clear that $\lim_{n \to \infty} \pi'(E_n) = \pi'(E)$ and $\lim_{n \to \infty} \pi''(E_n) = \pi''(E)$. But $\pi'(E_n) = \pi''(E_n)$, because $E_n \in \mathscr{S}$. Hence $\pi'(E) = \pi''(E)$; that is, $E \in \mathscr{S}$.
λ3. Let $E, F \in \mathscr{S}$ with $E \subset F$. Then $F = E \cup (F - E)$ is a disjoint union, so $\pi'(F) = \pi'(F - E) + \pi'(E)$ and $\pi''(F) = \pi''(F - E) + \pi''(E)$. Hence $\pi'(F - E) = \pi''(F - E)$, because $\pi'(F) = \pi''(F)$ and $\pi'(E) = \pi''(E)$.
b. Let $E = A \times B$ be a product set. Then $\phi_E(x) = v(B)1_A(x)$, so

$$\pi'(E) = \int_X \phi_E(x) \, d\mu = v(B)\mu(A)$$

Reversing the procedure gives

$$\pi''(E) = \mu(A)v(B)$$

Thus $E \in \mathscr{S}$.

Now, what happens if $\mu(X)$ and $v(Y)$ are not finite? Because X and Y are σ-finite, we can find subsets $X_i \in \mathcal{M}$, $i = 1, 2, 3, \ldots$, and $Y_j \in \mathcal{N}$, $j = 1, 2, 3, \ldots$, such that $\mu(X_i)$ and $v(Y_j)$ are finite and

$$X = \bigcup_{i=1}^{\infty} X_i \quad \text{(disjoint union)} \qquad \text{and} \qquad Y = \bigcup_{j=1}^{\infty} Y_j \quad \text{(disjoint union)}$$

If $E \in \mathcal{M} \times \mathcal{N}$ let

$$E_{i,j} = E \cap (X_i \times Y_j)$$

Then

(9)
$$E = \bigcup_{i,j=1}^{\infty} E_{i,j} \quad \text{(disjoint union)}$$

By what we proved above,

$$\pi'(E_{i,j}) = \pi''(E_{i,j})$$

so, by equation 9, $\pi'(E) = \pi''(E)$.   □

Definition 12. The measure $\pi' = \pi''$ is denoted $\mu \times \nu$ and is called the *product measure* on $\mathcal{M} \times \mathcal{N}$.

Example 13. Let $X = Y = \mathbf{R}$ and $\mathcal{M} = \mathcal{N} = \mathcal{B}_1$, the Borel sets in $\mathbf{R}$. Also let $\mu = \nu = \mu_1$, Lebesgue measure on $\mathbf{R}$. We claim that $\mathcal{M} \times \mathcal{N} = \mathcal{B}_2$, the Borel sets in $\mathbf{R}^2$, and $\mu \times \nu = \mu_2$, Lebesgue measure on $\mathbf{R}^2$.

Proof. Notice that, if I and J are intervals in $\mathbf{R}$, then $I \times J$ is a product set so $I \times J \in \mathcal{M} \times \mathcal{N}$. Now, $\mathcal{B}_2$ is the smallest σ-field containing sets of the form $I \times J$, so $\mathcal{B}_2 \subset \mathcal{M} \times \mathcal{N}$.

We now show $\mathcal{M} \times \mathcal{N} \subset \mathcal{B}_2$: Fix an interval $I \subset \mathbf{R}$. Let

$$\mathcal{B}_I = \{B \subset \mathbf{R}; I \times B \in \mathcal{B}_2\}$$

Note the following.

1. $\mathcal{B}_I$ contains all intervals $J \subset \mathbf{R}$.
2. $\mathcal{B}_I$ is a σ-field. Indeed, suppose $B_1, B_2, \ldots$ are elements of $\mathcal{B}_I$, then

$$I \times \left(\bigcup_{n=1}^{\infty} B_n \right) = \bigcup_{n=1}^{\infty} (I \times B_n) \in \mathcal{B}_2$$

so $\bigcup_{n=1}^{\infty} B_n \in \mathcal{B}_I$.

Because $\mathcal{B}_1$ is the smallest σ-field containing the intervals, items 1 and 2 imply that $\mathcal{B}_1 \subset \mathcal{B}_I$; that is, if $B \in \mathcal{B}_1$ we have $I \times B \in \mathcal{B}_2$.

Now fix a Borel set $B \in \mathcal{B}_1$. Let $\mathcal{B}_B = \{A \subset \mathbf{R}; A \times B \in \mathcal{B}_2\}$. Note the following.

1. $\mathcal{B}_B$ contains all intervals $I \subset \mathbf{R}$ by the above result.
2. $\mathcal{B}_B$ is a σ-field by an argument just like the one above.

Because $\mathcal{B}_1$ is the smallest σ-field containing all of the intervals, we have $\mathcal{B}_1 \subset \mathcal{B}_B$; that is, $A \times B \in \mathcal{B}_2$ for any $A, B \in \mathcal{B}_1$. But $\mathcal{B}_1 \times \mathcal{B}_1$ is the smallest σ-field containing all sets of the form $A \times B$, so we have shown that $\mathcal{B}_1 \times \mathcal{B}_1 \subset \mathcal{B}_2$ and thus $\mathcal{B}_1 \times \mathcal{B}_1 = \mathcal{B}_2$.

We now show that $\mu_1 \times \mu_1 = \mu_2$. Let

$$K_N = \{(x, y) \in \mathbf{R}^2;\ -N \le x, y \le N\}$$

It is enough to show that, for all N, $\mu_1 \times \mu_1 = \mu_2$ on Borel subsets of K_N. (Why?) To establish this fact, let $\mathcal{S}_N$ be the Borel subsets B of K_N for which

$$\mu_1 \times \mu_1(B) = \mu_2(B)$$

$\mathcal{S}_N$ is a λ-system:

λ1. $\mu_1 \times \mu_1(K_N) = 4N^2 = \mu_2(K_N)$.
λ2. Let $B_1 \subset B_2 \subset \cdots$ be an increasing sequence with each $B_n \in \mathcal{S}_N$. Let $B = \bigcup_{n=1}^{\infty} B_n$; then

$$(\mu_1 \times \mu_1)(B) = \lim_{n \to \infty} (\mu_1 \times \mu_1)(B_n) = \lim_{n \to \infty} \mu_2(B_n) = \mu_2(B)$$

λ3. Let E and F be elements of $\mathcal{S}_N$ with $E \subset F$. Then $F = E \cup (F - E)$, so

$$\mu_1 \times \mu_1(F - E) = \mu_1 \times \mu_1(F) - \mu_1 \times \mu_1(E)$$

$$= \mu_2(F) - \mu_2(E) = \mu_2(F - E)$$

Next, let $\mathcal{A}_N = \{I \times J;\ I, J \text{ subintervals of } [-N, N]\}$. $\mathcal{A}_N$ is a π-system because

$$(I_1 \times J_1) \cap (I_2 \times J_2) = (I_1 \cap I_2) \times (J_1 \cap J_2)$$

Furthermore, $\mathcal{A}_N \subset \mathcal{S}_N$ because

$$\mu_2(I \times J) = (\text{length } I) \cdot (\text{length } J)$$

whereas $\qquad (\mu_1 \times \mu_1)(I \times J) = \mu_1(I) \cdot \mu_1(J)$

$$= (\text{length } I) \cdot (\text{length } J)$$

From the $\pi-\lambda$ theorem we conclude that $\mathcal{S}_N$ contains the smallest σ-field containing $\mathcal{A}_N$. Hence $\mu_2 = \mu_1 \times \mu_1$ on all Borel subsets of K_N. $\qquad\square$

Exercise. Write $\mathbf{R}^3 = \mathbf{R}^2 \times \mathbf{R}^1$ and show that $\mathcal{B}_3 = \mathcal{B}_2 \times \mathcal{B}_1$ and $\mu_3 = \mu_2 \times \mu_1$.

We return now to the general situation: $(X, \mathcal{M}, \mu)$ and $(Y, \mathcal{N}, v)$ are σ-finite measure spaces and $(X \times Y, \mathcal{M} \times \mathcal{N}, \mu \times \nu)$ is the measure space constructed above.

Theorem 14. (Fubini, version 2) Let $f : X \times Y \to \mathbf{R}$ be a nonnegative measurable function. Then

1. **a.** For each $x \in X$, $f(x, y)$ is a measurable function of y.
    b. For each $y \in Y$, $f(x, y)$ is a measurable function of x.
2. **a.** $\int_Y f(x, y) \, d\nu$ is a measurable function of x.
    b. $\int_X f(x, y) \, d\mu$ is a measurable function of y.
3. $$\int_X \left[\int_Y f(x, y) \, d\nu \right] d\mu = \int_Y \left[\int_X f(x, y) \, d\mu \right] d\nu = \int_{X \times Y} f(x, y) \, d(\mu \times \nu)$$

Proof.   Part 1 was proved at the beginning of the section.

To prove parts 2 and 3, first note that they are true for $f(x, y) = 1_E(x, y)$ if $E \in \mathcal{M} \times \mathcal{N}$. Indeed

$$\int_Y 1_E(x, y) \, d\nu = \nu(E_x) = \phi_E(x)$$

which we have shown to be measurable, and similarly

$$\int_X 1_E(x, y) \, d\mu = \mu(E_y) = \psi_E(y)$$

is also measurable. The fact that

$$\int_X \left[\int_Y 1_E(x, y) \, d\nu \right] d\mu = \int_Y \left[\int_X 1_E(x, y) \, d\mu \right] d\nu = \int_{X \times Y} 1_E(x, y) \, d(\mu \times \nu)$$

is version 1 of the Fubini theorem.

Now, because parts 2 and 3 are true for characteristic functions, they must be true for simple functions by linearity. To prove the theorem in general, let $f(x, y)$ be a general measurable function and choose an increasing sequence of nonnegative simple functions

$$0 \le s_1 \le s_2 \le \cdots$$

with $s_n \to f$ pointwise. Then, for $x \in X$ fixed,

$$\int_Y s_n(x, y) \, d\nu \to \int_Y f(x, y) \, d\nu$$

by the monotone convergence theorem. Thus $\int_Y f(x, y) \, d\nu$ is a measurable function of x, because we know that $\int_Y s_n(x, y) \, d\nu$ is measurable. Similarly, $\int_X f(x, y) \, d\mu$ is a measurable function of y.

Furthermore, part 3 holds for all s_n; that is,

$$\int_X \left[\int_Y s_n(x, y)\, dv \right] d\mu = \int_Y \left[\int_X s_n(x, y)\, d\mu \right] dv = \int_{X \times Y} s_n(x, y)\, d(\mu \times v)$$

Applying the monotone convergence theorem to each of these terms separately yields

$$\int_X \left[\int_Y f(x, y)\, dv \right] d\mu = \int_Y \left[\int_X f(x, y)\, d\mu \right] dv = \int_{X \times Y} f(x, y)\, d(\mu \times v) \quad \square$$

Theorem 15.   (Fubini, version 3) Let f be integrable on $X \times Y$. Then

1. **a.**   For almost all x, $f(x, y)$ is integrable as a function of y.
    b.   For almost all y, $f(x, y)$ is integrable as a function of x.
2. **a.** $\int_Y f(x, y)\, dv$ is equal a.e. to an integrable function of x.
    b. $\int_X f(x, y)\, d\mu$ is equal a.e. to an integrable function of y.
3. $$\int_X \left[\int_Y f(x, y)\, dv \right] d\mu = \int_Y \left[\int_X f(x, y)\, d\mu \right] dv = \int_{X \times Y} f(x, y)\, d(\mu \times v).$$

Proof.   Write $f = f_+ - f_-$ where f_+ and f_- are nonnegative. Because f is integrable with respect to $\mu \times v$, we know that $\int_{X \times Y} f_+(x, y)\, d(\mu \times v)$ and $\int_{X \times Y} f_-(x, y)\, d(\mu \times v)$ are finite. Version 2 of Fubini then gives

$$\int_{X \times Y} f_+(x, y)\, d(\mu \times v) = \int_X \left[\int_Y f_+(x, y)\, dv \right] d\mu < \infty$$

Thus $\int_Y f_+(x, y)\, dv$ is finite a.e., and similarly $\int_X f_+(x, y)\, d\mu$ is finite a.e. This gives part 1. Parts 2 and 3 follow by applying version 2 of Fubini to f_+ and f_- separately and then adding. $\square$

A Final Remark.   Instead of considering only products of two measure spaces, we could have considered products of three or more measure spaces. For instance, let $(X_i, \mathcal{M}_i, \mu_i)$, $i = 1, 2, 3$, be measure spaces. One can define a product σ-field

$$\mathcal{M}_1 \times \mathcal{M}_2 \times \mathcal{M}_3$$

by defining it as either

$$(\mathcal{M}_1 \times \mathcal{M}_2) \times \mathcal{M}_3 \qquad \text{or} \qquad \mathcal{M}_1 \times (\mathcal{M}_2 \times \mathcal{M}_3)$$

or as the smallest σ-field containing the product sets

$$A_1 \times A_2 \times A_3$$

with $A_i \in \mathcal{M}_i$, $i = 1, 2, 3$. It turns out that these three definitions give the same σ-field. (See exercise 3.)

Moreover, we can define on $\mathcal{M}_1 \times \mathcal{M}_2 \times \mathcal{M}_3$ a product measure $\mu_1 \times \mu_2 \times \mu_3$ with the property that, on product sets,

(10) $$(\mu_1 \times \mu_2 \times \mu_3)(A_1 \times A_2 \times A_3) = \mu_1(A_1)\mu_2(A_2)\mu_3(A_3)$$

That is, we can define this measure as

$$\mu_1 \times (\mu_2 \times \mu_3) \quad \text{or} \quad (\mu_1 \times \mu_2) \times \mu_3$$

You will be asked in exercise 3 to show that these definitions are the same.

By simply using the Fubini theorem twice, we get analogous statements for this triple product. In particular, if f is integrable on $X_1 \times X_2 \times X_3$ (with respect to $\mu_1 \times \mu_2 \times \mu_3$), then the various partial integrals make sense a.e. and

(11) $$\int_{X_1 \times X_2 \times X_3} f\, d(\mu_1 \times \mu_2 \times \mu_3) = \int_{X_3} \left[\int_{X_2} \left(\int_{X_1} f\, d\mu_1 \right) d\mu_2 \right] d\mu_3$$

What we have said about products of three measure spaces applies equally well to products of any finite number of measure spaces. (See exercise 4.) Here the Fubini theorem looks like

(12) $$\int_{X_1 \times \cdots \times X_n} f\, d(\mu_1 \times \cdots \times \mu_n) = \int_{X_n} \cdots \left[\int_{X_1} f\, d\mu_1 \right] \cdots d\mu_n$$

Exercises for §2.5

1. Let R be the region

$$R = \{(x, y); \, -1 \leq x \leq 1, \, -1 \leq y \leq 1\}$$

in the plane. Compute the Lebesgue integrals

$$\int_R xy^2 \, d\mu_2 \qquad \int_R (x^2 + y^2) \, d\mu_2 \qquad \text{and} \qquad \int_R y e^{xy} \, d\mu_2$$

2. Let $(X, \mathcal{M}, \mu)$ and $(Y, \mathcal{N}, v)$ be σ-finite measure spaces. Call a product set $A \times B$ *finite* if $\mu(A) < \infty$ and $v(B) < \infty$. Show that the product measure $\mu \times v$ is the only measure satisfying

$$(\mu \times v)(A \times B) = \mu(A)v(B)$$

for all finite product sets $A \times B$. (*Hint:* Use the π–λ theorem.)

3. **a.** Let $(X_i, \mathcal{M}_i, \mu_i)$, $i = 1, 2, 3$, be σ-finite measure spaces. Show that

$$(\mathcal{M}_1 \times \mathcal{M}_2) \times \mathcal{M}_3 = \mathcal{M}_1 \times (\mathcal{M}_2 \times \mathcal{M}_3)$$

and that these equal the smallest σ-field containing the product sets $A_1 \times A_2 \times A_3$, where $A_i \in \mathcal{M}_i$, $i = 1, 2, 3$.

b. Show that $(\mu_1 \times \mu_2) \times \mu_3 = \mu_1 \times (\mu_2 \times \mu_3)$.

c. Call a product set $A_1 \times A_2 \times A_3$ *finite* if $\mu_i(A_i) < \infty$, $i = 1, 2, 3$. Show that the measure in part b is the only measure satisfying

$$(\mu_1 \times \mu_2 \times \mu_3)(A_1 \times A_2 \times A_3) = \mu_1(A_1)\mu_2(A_2)\mu_3(A_3)$$

for all finite product sets $A_1 \times A_2 \times A_3$.

4. Generalize exercise 3 to n-fold products.

5. In part a of exercise 3, let $X_1 = X_2 = X_3 = \mathbf{R}$, $\mathcal{M}_1 = \mathcal{M}_2 = \mathcal{M}_3 =$ the Borel sets of $\mathbf{R}$, and $\mu_1 = \mu_2 = \mu_3 =$ Lebesgue measure. Show that $\mathcal{M}_1 \times \mathcal{M}_2 \times \mathcal{M}_3$ is the Borel sets of $\mathbf{R}^3$ and $\mu_1 \times \mu_2 \times \mu_3$ is Lebesgue measure. Can you prove an equivalent statement for $\mathbf{R}^4$? for $\mathbf{R}^n$?

6. Let $(X, \mathcal{M}, \mu)$ and $(Y, \mathcal{N}, v)$ be σ-finite measure spaces, and let E be in $\mathcal{M} \times \mathcal{N}$. Show that E is of measure zero if and only if E_x is of measure zero for almost all $x \in X$.

7. **a.** Let $(X, \mathcal{M}, \mu)$ and $(Y, \mathcal{N}, v)$ be σ-finite measure spaces, and let $f : X \to \mathbf{R}$ and $g : Y \to \mathbf{R}$ be measurable functions. Let $h(x, y) = f(x)g(y)$. Show that h is a measurable function on $X \times Y$.

b. If f and g are integrable, show that h is integrable and that its integral is

$$\left(\int_X f \, d\mu \right)\left(\int_Y g \, dv \right)$$

8. (The integral as "area under the curve.") Let $(X, \mathcal{M}, \mu)$ be a σ-finite measure space, and let $f : X \to \mathbf{R}$ be a nonnegative measurable function. Let

$$A_f = \{(x, t) \in X \times \mathbf{R}; 0 \leq t \leq f(x)\}$$

Show that A_f is a measurable subset of $X \times \mathbf{R}$ (that is, belongs to $\mathcal{M} \times \mathcal{B}_1$) and that the measure of A_f with respect to the product measure $\mu \times \mu_{\text{Leb}}$ is equal to $\int_X f \, d\mu$.

9. Show that property $\lambda 2$ of a λ-system can be replaced by the property $\lambda 2'$. If $A_1, A_2, \ldots$ are disjoint subsets of $\mathcal{S}$, then $\bigcup_{i=1}^{\infty} A_i$ is in $\mathcal{S}$.

10. Let X be a set and $\mathcal{F}$ a σ-field of subsets of X. Let μ_1 and μ_2 be finite measures on $\mathcal{F}$. Show that the collection of sets

$$\{A \in \mathcal{F}; \mu_1(A) = \mu_2(A)\}$$

is a λ-system.

11. Show that Lebesgue measure is the only measure on the Borel sets of the interval $[0, 1]$ with the property that, for all subintervals J, $\mu(J) = $ length of J. (*Hint:* Use exercise 10 and the $\pi{-}\lambda$ theorem.)

12. a. Show that for the function

$$f(x, y) = \frac{xy}{(x^2 + y^2)^2}$$

the iterated integrals

$$\int_{-1}^{1} \left[\int_{-1}^{1} f(x, y)\, dy \right] dx \quad \text{and} \quad \int_{-1}^{1} \left[\int_{-1}^{1} f(x, y)\, dx \right] dy$$

exist and are equal.

b. Show that f is not integrable over the square $-1 \leq x \leq 1$ and $-1 \leq y \leq 1$.

13. Let $X = Y = \mathbf{R}$ and $\mathcal{M} = \mathcal{N} = $ Borel sets. Let μ be Lebesgue measure, and let v be counting measure; that is,

$$v(B) = \text{number of elements in } B$$

Let $E \in \mathcal{M} \times \mathcal{N}$ be the set

$$E = \{(x, y) \in X \times Y;\ x = y\}$$

Recall that $\phi_E(x) = v(E_x)$ and $\psi_E(y) = \mu(E_y)$. Show that ϕ_E and ψ_E are measurable but that

$$\int_X \phi_E(x)\, d\mu \neq \int_Y \psi_E(y)\, dv$$

§2.6 Random Variables, Expectation Values, and Independence

In the next two sections we will discuss some probabilistic applications of the material in §2.5. Let X be a set, $\mathcal{F}$ a σ-field of subsets of X, and μ a probability measure on X. A *random variable f* is, by definition, a measurable function $f : X \to \mathbf{R} \cup \{\pm\infty\}$. For instance, let $X = I = \mathcal{B}$ and let

(1)
$$f = \frac{1}{2}\left(n + \sum_{k=1}^{n} R_k\right)$$

Interpreted probabilistically, f is the number of times H comes up in the first n stages of a Bernoulli sequence. It is a "random quantity" or "random variable" that can be measured each time we perform a sequence of Bernoulli trials.

The *expectation value* of a random variable is its integral

$$E(f) = \int_X f\, d\mu$$

Of course, the expectation value need not always be defined; that is, f need not always be integrable. However, all the random variables we will consider do have well-defined expectation values. For instance, if we integrate equation 1 over the unit interval we get

$$\int_X f \, d\mu = E(f) = \frac{n}{2}$$

representing the fact that $n/2$ is the number of heads "most likely" to turn up in a sequence of n Bernoulli trials. (This meaning of expectation value will become clearer in §2.7.)

Given a random variable $f : X \to \mathbf{R}$ and a Borel subset $A \subseteq \mathbf{R}$ let

(2) $$\mu_f(A) = \mu[f^{-1}(A)]$$

The right-hand quantity is well defined because $f^{-1}(A) \in \mathscr{F}$. (See §2.1.) We leave it for the reader to check that equation 2 defines a measure on the Borel subsets of $\mathbf{R}$. (See §2.3, exercise 15.) We will call this measure the *probability distribution* associated with the random variable f. If, for two random variables f and g, $\mu_f = \mu_g$, we will say that f and g are *identically distributed*. The essential property for us of the measure μ_f is the following.

Theorem 1. Let ϕ be a nonnegative Borel-measurable function on $\mathbf{R}$. Then

(3) $$\int_X \phi(f) \, d\mu = \int_{\mathbf{R}} \phi \, d\mu_f$$

Proof. First, suppose that ϕ is the characteristic function of a Borel subset $A \subseteq \mathbf{R}$. Then $\phi(f)$ is the characteristic function of $f^{-1}(A)$, so the left-hand side of equation 3 is $\mu[f^{-1}(A)]$ and the right-hand side is $\mu_f(A)$. By equation 2 these quantities are equal. Next observe that equation 3 holds for finite linear combinations of characteristic functions of sets—that is, for *simple* functions. Finally, by theorem 6 of §2.2 there exists an increasing sequence s_n of nonnegative simple functions with $s_n \to \phi$. Then $s_n(f) \to \phi(f)$; so equation 3 follows from the monotone convergence theorem. ☐

Corollary 2. Let ϕ be a Borel-measurable function on $\mathbf{R}$. Then ϕ is integrable with respect to the measure μ_f if and only if $\phi(f)$ is integrable with respect to μ. When such is the case, equation 3 holds.

Proof. Let $\phi = \phi_+ - \phi_-$ and apply theorem 1 to ϕ_+ and ϕ_- separately. ☐
Notice that if $\phi(x) = x$, then equation 3 becomes

$$\int_{\mathbf{R}} x \, d\mu_f = \int_X f \, d\mu = E(f)$$

This shows that, *if random variables f and g are identically distributed, they have the same expectation value.* More generally, if f and g are identically distributed, then, for every Borel-measurable function ϕ on $\mathbf{R}$,

$$\int \phi \, d\mu_f = \int \phi \, d\mu_g$$

so by equation 3

(4)
$$\int_X \phi(f) \, d\mu = \int_X \phi(g) \, d\mu$$

For instance, taking $\phi(x) = x^2$ we get

$$\int f^2 \, d\mu = \int g^2 \, d\mu$$

Given several random variables $f_1, \dots, f_n$, let $F : X \to \mathbf{R}^n$ be the map $F(x) = (f_1(x), \dots, f_n(x))$. If A is a Borel subset of $\mathbf{R}^n$, set

(5)
$$\mu_{f_1, \dots, f_n}(A) = \mu[F^{-1}(A)]$$

This formula defines a probability measure on the Borel subsets of $\mathbf{R}^n$ (check this!) called the *joint probability distribution associated with* $f_1, \dots, f_n$. The analogue of equation 3,

(6)
$$\int_X \phi(f_1, \dots, f_n) \, d\mu = \int_{\mathbf{R}} \phi \, d\mu_{f_1, \dots, f_n}$$

holds for any nonnegative Borel-measurable function ϕ on $\mathbf{R}^n$ and is proved in exactly the same way.

A set of random variables $f_1, \dots, f_n$ is said to be *independent* if, for any sequence of Borel subsets $A_1, A_2, \dots, A_n$ of $\mathbf{R}$, the sets

$$f_1^{-1}(A_1), \dots, f_n^{-1}(A_n)$$

are independent as subsets of X. An infinite sequence of random variables $f_1, f_2, \dots$ is said to be independent if every finite subsequence is independent. A very simple criterion for independence in terms of the joint probability distributions of the f_i's is the following.

Theorem 3.   The random variables $f_1, \dots, f_n$ are independent if and only if the probability measure $\mu_{f_1, \dots, f_n}$ is equal to the product measure $\mu_{f_1} \times \mu_{f_2} \times \cdots \times \mu_{f_n}$.

Remark.   The product is, of course, defined as in §2.5.

Proof. To check that the measures agree, it is enough, by the $\pi-\lambda$ theorem, to check that they agree on sets of the form $A_1 \times \cdots \times A_n$. By definition

$$\mu_{f_1 \times \cdots \times f_n}(A_1 \times \cdots \times A_n) = \mu[f_1^{-1}(A_1) \cap \cdots \cap f_n^{-1}(A_n)]$$

and by independence the right-hand expression is equal to $\mu[f_1^{-1}(A_1)] \times \cdots \times \mu[f_n^{-1}(A_n)]$ or $\mu_{f_1}(A_1) \times \cdots \times \mu_{f_n}(A_n)$ or $(\mu_{f_1} \times \cdots \times \mu_{f_n})(A_1 \times \cdots \times A_n)$. $\square$

One consequence of theorem 3 is the identity

(7)
$$\int f_1 \times \cdots \times f_n \, d\mu = E(f_1) \times \cdots \times E(f_n)$$

Indeed, if we apply equation 5 to the function $\phi(x_1, \ldots, x_n) = x_1 \cdots x_n$ we get

$$\int_X f_1 \times \cdots \times f_n \, d\mu = \int_{\mathbf{R}^n} x_1 x_2 \cdots x_n \, d\mu_{f_1, \ldots, f_n}$$

$$= \int_{\mathbf{R}^n} x_1 x_2 \cdots x_n \, d\mu_{f_1} \times \cdots \times d\mu_{f_n}$$

$$= \left(\int_{\mathbf{R}} x_1 \, d\mu_{f_1} \right) \left(\int_{\mathbf{R}} x_2 \, d\mu_{f_2} \right) \cdots \left(\int_{\mathbf{R}} x_n \, d\mu_{f_n} \right)$$

by Fubini's theorem. However, the ith term on the right is just $E(f_i)$.

A remark about independence that will be useful below is the following. Let $f_1, \ldots, f_n$ be independent, and let $\phi_1, \ldots, \phi_n$ be Borel-measurable functions on $\mathbf{R}$. Then

(8)
$$\phi_1(f_1), \ldots, \phi_n(f_n)$$

are independent. In fact, let $A_1, \ldots, A_n$ be Borel subsets of $\mathbf{R}$, and let $A_i' = \phi_i^{-1}(A_i)$. Then the A_i''s are also Borel subsets, and

$$[\phi_i(f_i)]^{-1}(A_i) = f_i^{-1}(A_i') \qquad i = 1, \ldots, n$$

By assumption, the sets on the right are independent; hence, so are the sets on the left.

Example 4. If f_1 and f_2 are independent, then f_1^3 and $|f_2|$ are independent.

The notion of independence plays a central role in measure-theoretic models of probabilistic processes. For instance let's go back to the gambling process described at the end of §1.2. Recall that this process involves a cage filled with colored marbles. There are assumed to be k different colors, with N_i marbles of each color i and $N = \sum_{i=1}^{k} N_i$ marbles in all. The process consists of mixing the marbles, then drawing a marble out of the cage. If the color of the marble is i, the player receives a reward (or penalty) of r_i dollars. The marble is then replaced, the marbles are again thoroughly mixed, another marble is drawn, and the game continues. If f_n is the amount of the reward or penalty at the nth draw, then f_n takes on the values of $r_1,\ldots,r_k$ with the probabilities of $p_1,\ldots,p_k$, where $p_i = N_i/N$. What is an adequate measure-theoretic description of this situation? We claim that the data needed to "model" this process are: (a) a set X, a σ-field $\mathscr{F}$ of subsets of X, and a probability measure μ on $\mathscr{F}$; and (b) an infinite sequence of random variables $f_1, f_2,\ldots$ with the following properties:

(9)   The f_i's are independent

and

(10)
$$\mu_{f_i}(A) = \sum_{r_j \in A} p_j$$

for every Borel subset A of $\mathbf{R}$. Indeed, the f_i's give an identification of points $x \in X$ with infinite sequences of draws from the cage, i.e., $f_k(x)$ describes what happens in the sequence corresponding to x at the kth draw. Property 9 just says that what happens at the nth draw is independent of what happens at any of the other draws. This is justified by the fact that the marbles are thoroughly mixed after each draw. If one sets $A = \{r_m\}$, then, by equation 10, the proba-bility that at stage n the reward or penalty incurred will be r_m is just

$$\mu_{f_n}(A) = p_m$$

which is what we expect because of the number of marbles of color m in the cage. Notice that equation 10 implies that the f_i's are identically distributed.

We will now show that a probabilistic model with all the above features does exist. In fact, we will show that we can even take for X the unit interval I, for $\mathscr{F}$ the Borel subsets of I, and for the probability measure on $\mathscr{F}$ ordinary Lebesgue measure.

Theorem 5.   There exist bounded measurable functions $f_1, f_2,\ldots$ on I such that property 9 and equation 10 hold with $\mu = \mu_L =$ Lebesgue measure.

Proof.   Decompose the unit interval into k disjoint subintervals $I_1,\ldots,I_k$ such that I_l is of length p_l, and define f_1 by setting

$$f_1 = r_l \quad \text{on } I_l \qquad l = 1,\ldots,n$$

Next decompose each of the intervals I_l into k disjoint intervals $I_{l,m}, m = 1,\ldots, k$, such that $I_{l,m}$ is of length $p_l p_m$. (Because I_l is of length p_l and $\sum p_m = 1$, such a choice of $I_{l,m}$'s is clearly possible.) Define f_2 by setting

$$f_2 = r_m \quad \text{on } I_{l,m}$$

Notice that, if $A_l = \{r_l\}$ and $A_m = \{r_m\}$,

$$f_1^{-1}(A_l) \cap f_2^{-1}(A_m) = I_{l,m}$$

so

$$\mu[f_1^{-1}(A_l) \cap f_2^{-1}(A_m)] = p_l p_m$$

On the other hand,

$$f_1^{-1}(A_l) = I_l \quad \text{and} \quad f_2^{-1}(A_m) = \bigcup_{l=1}^{k} I_{l,m} \quad \text{(disjoint union)}$$

so

$$\mu[f_1^{-1}(A_l)] = p_l \quad \text{and} \quad \mu[f_2^{-1}(A_m)] = \sum_{l=1}^{k} p_l p_m = p_m$$

From these computations we conclude that

$$\mu[f_1^{-1}(A_l) \cap f_2^{-1}(A_m)] = \mu[f_1^{-1}(A_l)]\mu[f_2^{-1}(A_m)]$$

or, in other words, $f_1^{-1}(A_l)$ and $f_2^{-1}(A_m)$ are independent. Now, suppose that B_1 and B_2 are arbitrary Borel subsets of $\mathbf{R}$. Then

$$B_1 = B_1' \cup \left(\bigcup_{r_l \in B_1} A_l \right) \quad \text{and} \quad B_2 = B_2' \cup \left(\bigcup_{r_m \in B_2} A_m \right)$$

with B_1' and B_2' containing none of the r_l's. Then $f_1^{-1}(B_1')$ and $f_2^{-1}(B_2')$ are empty; so

$$f_1^{-1}(B_1) = \bigcup_{r_l \in B_1} f_1^{-1}(A_l) \quad \text{(disjoint union)}$$

and

$$f_2^{-1}(B_2) = \bigcup_{r_m \in B_2} f_2^{-1}(A_m) \quad \text{(disjoint union)}$$

are independent. (Why?) Because B_1 and B_2 are arbitrary, f_1 and f_2 are independent. We let the reader check that f_1 and f_2 satisfy equation 10 and move on to the construction of f_3. Decompose $I_{l,m}$ into k disjoint subintervals $I_{l,m,n}$, $n = 1, \ldots, k$, of length $p_l p_m p_n$ and define f_3 by setting

$$f_3 = r_n \quad \text{on } I_{l,m,n}$$

One checks that f_1, f_2, and f_3 are independent in exactly the same way as above. It is also clear by now how to construct f_4, f_5, and so on. We leave details to the reader. $\qquad \square$

Remark. Let $k = 2, r_1 = 1, r_2 = -1, p_1 = \frac{1}{2}$, and $p_2 = \frac{1}{2}$. Then the functions constructed above are exactly the Rademacher functions!

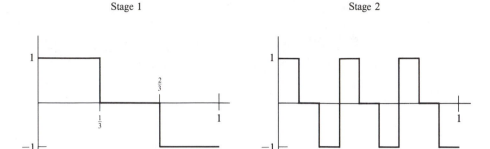

This figure indicates the first two stages in the construction of the f_i's with the data $k = 3, r_1 = 1$, $r_2 = 0, r_3 = -1$, and $p_1 = p_2 = p_3 = \frac{1}{3}$.

Exercises for §2.6

1. Let v be a probability measure on the Borel sets of the real line. Then v is said to be supported on the Borel set A if $\mu(A^c) = 0$. Using theorem 1, show that, if A is a Borel set containing the image of f, then μ_f is supported on A. In particular, if f is a simple function taking on the values $r_1, \ldots, r_k$, then μ_f is supported on $\{r_1, \ldots, r_k\}$.

2. Let $X = I$ be the unit interval, and let μ be Lebesgue measure. Describe the measure μ_f for the function $f(x) = x^2$.

3. If, for $i = 1, 2, f_i = R_i$ is the ith Rademacher function, what is the measure μ_{f_i}? Verify directly that $\mu_{f_1, f_2} = \mu_{f_1} \times \mu_{f_2}$.

4. (The unfair coin.) Using theorem 5, let $k = 2, r_1 = 1, r_2 = -1, p_1 = p$, and $p_2 = 1 - p$. Describe the first three of the functions $f_1, f_2, f_3, \ldots$.

5. In exercise 4, let $S_n = f_1 + \cdots + f_n$. Compute the expectation value E of S_n and the variance

$$V(S_n) = \int (S_n - E)^2 \, d\mu_L$$

6. (The random walk with pauses.) Using theorem 5, let $k = 3, r_1 = 1, r_2 = 0$, $r_3 = -1$, and $p_1 = p_2 = p_3 = \frac{1}{3}$. We have already drawn the graphs of f_1 and f_2. Draw the graph of f_3. Can you discern a pattern?

7. **a.** Let R_i be the ith Rademacher function, and let

$$f_k(x) = \sum_{i=1}^{k} \left(\frac{1}{2^i}\right) R_i(x)$$

Compute

$$\int_I e^{t f_k(x)} \, dx$$

(*Hint:* Use independence.)

b. Using the formula

$$2x - 1 = \lim_{k \to \infty} f_k(x)$$

(proved in exercise 7 of §1.1) deduce Vieta's formula

$$\frac{\sinh t}{t} = \prod_{k=1}^{\infty} \cosh\left(\frac{t}{2^k}\right)$$

8. a. Let $f_1, f_2, \ldots$ be as in theorem 5, and let $S_n = f_1 + \cdots + f_n$. Prove that

$$\int_I e^{tS_n(x)} \, dx = \sum_r \text{Prob}(S_n = r)e^{tr}$$

b. Conclude from part a that

(*) $$\sum_r \text{Prob}(S_n = r)e^{tr} = \left(\sum_{i=1}^{k} p_i e^{tr_i}\right)^n$$

(*Hint:* Write the integral on the left in part a as

$$\int_I e^{tf_1} e^{tf_2} \times \cdots \times e^{tf_n} \, dx$$

and use independence.)

9. Let $f_1, f_2, f_3, \ldots$ be independent, identically distributed random variables taking on the value of 1 with probability p and the value of 0 with probability $1 - p$, where $0 \le p \le 1$. (That is, using theorem 5, take $k = 2$, $r_1 = 1, r_2 = 0, p_1 = p$, and $p_2 = 1 - p$.) Let $S_n = f_1 + \cdots + f_n$. Show that

(**) $$\text{Prob}(S_n = r) = \binom{n}{r} p^r (1 - p)^{n-r}$$

if $0 \le r \le n$ and is zero otherwise. (*Hint:* Use exercise 8.)

10. Let $\{r_1, r_2, \ldots\}$ be a countable subset of $\mathbf{R}$, and let $p_1, p_2, \ldots$ be a countable sequence of nonnegative numbers with

$$\sum_{i=1}^{\infty} p_i = 1$$

For every subset A of $\mathbf{R}$, let

(†) $$v(A) = \sum_{r_i \in A} p_i$$

Show that there exists a sequence $f_1, f_2, \ldots$ of independent, identically distributed random variables on the unit interval such that $\mu_{f_1} = \mu_{f_2} = \cdots = v$.

11. For $i = 1, \ldots, n$ let $\{X_i, \mathcal{M}_i, \mu_i\}$ be a probability space, let $\mathcal{M}_1 \times \cdots \times \mathcal{M}_n$ be the product of the $\mathcal{M}_i$'s, and let $\mu_1 \times \cdots \times \mu_n$ be the product measure on $\mathcal{M}_1 \times \cdots \times \mathcal{M}_n$. For each i let f_i be a measurable function on X_i. Consider f_i as a function on $X_1 \times \cdots \times X_n$ by setting

$$f_i(x_1, \ldots, x_n) = f_i(x_i)$$

Show that the f_i's, regarded as functions on $X_1 \times \cdots \times X_n$, are independent.

§2.7 The Law of Large Numbers

Let's now return to the question posed at the beginning of §2.6. What does the expectation value of a random variable really represent? Consider a simple probabilistic process (such as the toss of a coin) and a numerical quantity Q associated with the process. (For instance, for the toss of a coin let $Q = 1$ if an H occurs and $Q = -1$ if a T occurs.) Now repeat the process and again measure the quantity Q; repeat it a third time and again measure Q, and so on ad infinitum. Let $AV_n(Q)$ be the average value of Q, averaged over the first n stages of this infinite sequence of experiments. Does $AV_n(Q)$ tend to a limit as n tends to infinity? The answer is yes, provided that, each time the experiment is repeated, the conditions under which it is performed are not biased by the results of the preceding trials. (For instance, if the experiment consists of drawing a marble from a cage, recording its color, and then replacing, it, the marbles must be thoroughly mixed each time.) We will show that, if these experimental requirements are met, then, not only does $AV_n(Q)$ tend to a limit as $n \to \infty$, but in fact

(1) $$\lim_{n \to \infty} AV_n(Q) = E(Q)$$

is the expectation value of Q. (We mean, of course, that equation (1) holds with probability one.) To see this, let's first describe the experimental set-up above in somewhat more precise terms. Let f_n be the measured value of the quantity Q at the nth stage of the sequence of experiments. Then, under the hypotheses above, the f_n's are *independent, identically distributed random variables*. The underlying space X on which they are defined is, technically speaking, the totality of "all infinite sequences of repetitions of the experiment." For instance, if the experiment consists of the toss of a coin, X is the set $\mathcal{B}$ of all Bernoulli sequences as in §1.1. Actually it isn't terribly important to describe X this explicitly. What is important are the i.i.d. (independent, identically distributed) random variables $f_1, f_2, \ldots$ and their common probability distribution $\mu_{f_1} = \mu_{f_2} = \cdots$. For instance, for the experiment described in §2.6 (a colored marble drawn from a cage) we showed that X could be taken to be a very simple set: the unit interval. The important point was that on the unit interval we could produce a sequence of independent random variables $f_1, f_2, \ldots$ all with the probability distribution of equation 10 in §2.6.

The following result is what is traditionally called the law of large numbers. Fix a set X, a σ-field $\mathscr{F}$ of subsets of X, and a probability measure μ on $\mathscr{F}$.

Theorem 1. Let $f_1, f_2, \ldots$ be a sequence of bounded random variables on X that are independent and identically distributed. Let $E = E(f_1) = E(f_2) = \cdots$ be the common expectation value of the f_i's. Let X_0 be the set of points $x \in X$ for which

$$(2) \qquad \frac{f_1(x) + \cdots + f_n(x)}{n} \to E$$

as $n \to \infty$. Then $\mu(X_0) = 1$.

Remark. The assumption that the f_i's are bounded is not essential, but it simplifies some details of the proof.

Proof. Replacing f_i by $f_i - E$, we can, without loss of generality, assume that $E = 0$. Let

$$V = \left(\int f_i^2 \, d\mu \right)^2 \qquad \text{and} \qquad W = \int f_i^4 \, d\mu$$

Because the f_i's are identically distributed, these quantities are the same for all i's. The first step in the proof will be to establish the following inequality for all $\varepsilon > 0$:

$$(3) \qquad \mu\left(\left\{ x \in X; \left| \frac{f_1(x) + \cdots + f_n(x)}{n} \right| > \varepsilon \right\} \right) \leq \frac{3n(n-1)V + nW}{\varepsilon^4 n^4}$$

The left-hand side of inequality 3 is equal to

$$\mu(\{x \in X; (f_1 + \cdots + f_n)^4 \geq n^4 \varepsilon^4\})$$

and, by Chebyshev's inequality, this is less than

$$\left(\frac{1}{n^4 \varepsilon^4} \right) \int (f_1 + \cdots + f_n)^4 \, d\mu$$

so inequality 3 reduces to

$$(4) \qquad \int (f_1 + \cdots + f_n)^4 \, d\mu \leq 3n(n-1)V + nW$$

If we multiply out the expression on the left, we get four sorts of terms— namely,

$$\int f_\alpha^4 \, d\mu$$

$$\int f_\alpha^2 f_\beta^2 \, d\mu \qquad \alpha \neq \beta$$

$$\int f_\alpha^2 f_\beta f_\gamma \, d\mu \qquad \alpha \neq \beta \neq \gamma$$

and
$$\int f_\alpha f_\beta f_\gamma f_\delta \, d\mu \qquad \alpha \neq \beta \neq \gamma \neq \delta$$

The first integral is equal to W, and the second integral is equal to

$$\left(\int f_\alpha^2 \, d\mu \right) \left(\int f_\beta^2 \, d\mu \right) \qquad \text{or} \qquad V$$

by equation 7 of §2.6. Similarly, the third integral is equal to

$$\left(\int f_\alpha^2 \, d\mu \right) \left(\int f_\beta \, d\mu \right) \left(\int f_\gamma \, d\mu \right)$$

and the fourth integral is equal to

$$\left(\int f_\alpha \, d\mu \right) \left(\int f_\beta \, d\mu \right) \left(\int f_\gamma \, d\mu \right) \left(\int f_\delta \, d\mu \right)$$

Because the expectation values are zero, both these terms are zero. Because there are exactly n integrals of the first type and $3n(n-1)$ integrals of the second type (see §1.1), the sum of all these integrals is the right-hand side of inequality 4.

Now choose a sequence of numbers $\varepsilon_1, \varepsilon_2, \varepsilon_3, \ldots$ such that $\varepsilon_n \to 0$ and

$$\sum_{n=1}^{\infty} \frac{3n(n-1)V + nW}{\varepsilon_n^4 n^4} < \infty$$

(See lemma 6 in §1.1.) Let

$$A_n = \left\{ x \in X; \; \left| \frac{f_1(x) + \cdots + f_n(x)}{n} \right| > \varepsilon_n \right\}$$

Then, by inequality 3, $\sum_{n=1}^{\infty} \mu(A_n) < \infty$. So, by the first Borel–Cantelli lemma, $\mu(A_n; \text{i.o.}) = 0$. This result means that, if we exclude a set of measure zero from X, then for x in the complement

$$\left| \frac{f_1(x) + \cdots + f_n(x)}{n} \right| < \varepsilon_n$$

for all but finitely many values of n. This clearly implies that

$$\frac{f_1(x) + \cdots + f_n(x)}{n} \to 0 \quad \text{as} \quad n \to \infty \qquad \square$$

Exercises for §2.7

1. Show that, if the f_n's in theorem 1 are the Rademacher functions, then equation 2 is the strong law of large numbers formulated in §1.1.
2. Let $(X, \mathscr{F}, \mu)$ be a probability space. Recall (§2.3, exercise 8) that a sequence of measurable functions f_n, $n = 1, 2, \ldots$ *converges to zero in measure* if for all $\varepsilon > 0$

$$\mu(\{x \in X; |f_n(x)| > \varepsilon\}) \to 0 \quad \text{as} \quad n \to \infty$$

Show that, if $\{f_n\}$ converges pointwise to zero almost everywhere, then f_n converges to zero in measure. (*Hint:* Let $A_n = \{x \in X; |f_k(x)| < \varepsilon \text{ for } k \geq n\}$. Show that $A_1 \subseteq A_2 \subseteq A_3 \subseteq \cdots$ and that $X - \bigcup_{n=1}^{\infty} A_n$ is of measure zero.)
3. Deduce from exercise 2 and theorem 1 the *weak law of large numbers*. Show that, if $f_1, f_2, \ldots$ are a sequence of bounded independent, identically distributed random variables and if E is their common expectation value, then for all $\varepsilon > 0$

$$\text{Prob}\left(\left|\frac{f_1 + \cdots + f_n}{n} - E\right| > \varepsilon\right) \to 0 \quad \text{as} \quad n \to \infty$$

4. Show that, if $f_1, f_2, \ldots$ are as in exercise 3, then

$$\text{Prob}\left(\left|\frac{f_1 + \cdots + f_n}{n} - E\right| > \varepsilon\right) \leq \left(\frac{1}{\varepsilon^2 n}\right) V$$

where V is the common variance of the f_i's, i.e., $V = \int (f_i - E)^2 \, d\mu$. Use this to give another proof of the weak law of large numbers.
5. Let $f_1, f_2, \ldots$ be bounded, independent, identically distributed random variables with $E = E(f_1) = E(f_2) = \cdots = 0$. Let $S_n = f_1 + \cdots + f_n$. Show that if $\alpha > 0$ then

$$S_n(x)/n^{(1/2)+\alpha} \to 0 \text{ a.e.} \quad \text{as} \quad n \to \infty$$

(Compare with exercise 17 in §1.1.) (*Hint:* Show that there exists a constant C_k such that

$$\int S_n^{2k} \, d\mu \leq C_k n^k$$

for every integer $k > 0$.)

6. Let $f_1, f_2, \ldots$ be independent, identically distributed random variables on the unit interval. Suppose that the common probability distribution of the f_i's is given by equation 9 of §2.6, with $k = 2$, $r_1 = 1$, $r_2 = 0$, $p_1 = p$, and $p_2 = 1 - p$ (p being any number between zero and one). Let $S_n = f_1 + \cdots + f_n$. Show that, if ϕ is any bounded measurable function on the interval $[0, 1]$,

(*) $$E\left[\phi\left(\frac{S_n}{n}\right)\right] = \int \phi\left(\frac{S_n}{n}\right) d\mu = \sum_{k=0}^{n} \phi\left(\frac{k}{n}\right)\binom{n}{k} p^k (1 - p)^{n-k}$$

(*Hint:* See section 2.6, exercise 9.)

Remark.   We will denote the right-hand side of equation (*) by $B_n(\phi, p)$. Notice that it is a polynomial of degree n in p. We will call it the nth *Bernstein polynomial associated with ϕ.*

7. Show that for n very large $S_n(x)/n$ is very close to p for most values of x. Explicitly show that for all $\delta > 0$

(**) $$\mu\left(\left\{x \in [0, 1]; \left|\frac{S_n(x)}{n} - p\right| > \delta\right\}\right) \le \left(\frac{1}{\delta^2 n}\right) p(1 - p)$$

(*Hint:* See exercise 4.)

8. Let ϕ be a continuous function on the interval $[0, 1]$. Show that

(†) $$E\left[\phi\left(\frac{S_n}{n}\right)\right] \to \phi(p) \quad \text{as} \quad n \to \infty$$

(*Hint:* Given $\varepsilon > 0$ choose δ so that $|\phi(s) - \phi(t)| < \varepsilon$ when $0 \le s, t \le 1$ and $|s - t| < \delta$. Let I_1 be the subset of $0 \le x \le 1$ on which $|(S_n(x)/n) - p| < \delta$, and let I_2 be the complementary set. Show that

$$\int_{I_1}\left|\phi\left(\frac{S_n(x)}{n}\right) - \phi(p)\right| dx < \varepsilon$$

and estimate

$$\int_{I_2}\left|\phi\left(\frac{S_n(x)}{n}\right) - \phi(p)\right| dx$$

using inequality (**) and the fact that ϕ is bounded.)

9. Show that the convergence in display (†) is uniform in p. By equation (*) conclude that, as functions of p, the Bernstein polynomials $B_n(\phi)$ converge uniformly to ϕ on the interval $[0, 1]$. (The result we have asked you to prove is a constructive form of the *Weierstrass approximation theorem:* Given a continuous function ϕ on the interval $[0, 1]$, there exists a sequence of polynomials B_n converging uniformly to ϕ as $n \to \infty$.)

§2.8 The Discrete Dirichlet Problem

Let $\mathcal{O}$ be an open set in $\mathbf{R}^2$. A twice-differentiable function $f : \mathcal{O} \to \mathbf{R}$ is called *harmonic* on $\mathcal{O}$ if

$$\frac{\partial^2 f}{\partial x^2} + \frac{\partial^2 f}{\partial y^2} = 0 \qquad \text{on } \mathcal{O}$$

Now let Ω be a compact subset of $\mathbf{R}^2$ with a continuous boundary $\partial\Omega$. Suppose that $g : \partial\Omega \to \mathbf{R}$ is continuous. The classical *Dirichlet problem* asks one to find $f : \Omega \to \mathbf{R}$ such that f is harmonic on $\text{Int}\,\Omega$ and $f = g$ on $\partial\Omega$.

Many solutions to this problem have been discovered, some of which are quite ingenious. In particular, in *Two dimensional Brownian motion and harmonic functions* (Tokyo: Proc. Imp. Acad., 20, 706–714 [1944]), S. Kakutani showed how to construct f using probabilistic methods. He used a kind of limit of the random walk in $\mathbf{R}^2$ called the Wiener process or Brownian motion. Although the theory of the Wiener process is beyond the scope of this book, we can understand the ideas behind Kakutani's construction by looking at a discrete version of the Dirichlet problem due to Courant (Courant, R., Friedrichs, K. O., and Lewy, H. *Ueber die partiellen Differenzengleichungen der mathematischen Physik.* Math. Ann. Vol. 100. pp. 32–74 [1928]). (In fact, Courant showed that the solution to the classical problem can be obtained as a limiting case of the solution of the discrete problem described below!)

Before we describe this discrete version of the Dirichlet problem, we need to translate the definition of harmonic functions into a form that is easily dealt with measure theoretically.

Theorem.   (Mean value property) Let $\mathcal{O} \subset \mathbf{R}^2$ be open and let $f : \mathcal{O} \to \mathbf{R}$ be harmonic. Let $x_0 \in \mathcal{O}$ and assume that the circle of radius a around x_0 lies entirely in $\mathcal{O}$. Then

(1) $$f(x_0) = \left(\frac{1}{2\pi}\right) \int_0^{2\pi} f(x_0 + ae^{i\theta})\, d\theta$$

Conversely, if $f : \mathcal{O} \to \mathbf{R}$ is continuous and equation 1 holds for all x_0 and a such that the circle of radius a around x_0 lies entirely in $\mathcal{O}$, then f is twice-differentiable and harmonic in $\mathcal{O}$.

For a proof of this theorem see, for example, L. Ahlfors, *Complex Analysis* (New York: McGraw-Hill [1953]).

Using this characterization of harmonic functions, we can formulate a plausible discrete analogue of the Dirichlet problem. The space $\mathbf{R}^2$ is replaced by the integer lattice

$$\mathbf{Z}^2 = \{(m, n);\ m, n \text{ are integers}\}$$

and the compact region Ω becomes a finite subset of $\mathbf{Z}^2$.

For $x \in \mathbf{Z}^2$, there are four nearest neighbors, x_N, x_S, x_E, and x_W, as pictured below.

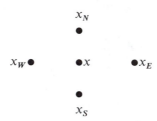

If $x \in \Omega$ we say $x \in \text{Int } \Omega$ if x_N, x_S, x_E, and x_W are all in Ω as well. We then define $\partial \Omega = \Omega - \text{Int } \Omega$.

To define harmonic functions on $\text{Int } \Omega$, the integral in equation 1 is translated to be the average over the nearest neighbors. Namely, if $f : \Omega \to \mathbf{R}$ we say f is *harmonic* on $\text{Int } \Omega$ if

$$f(x) = \frac{1}{4}[f(x_N) + f(x_S) + f(x_E) + f(x_W)]$$

for all $x \in \text{Int } \Omega$.

Now let's consider the following problem.

Discrete Dirichlet Problem

Given $g : \partial \Omega \to \mathbf{R}$ find $f : \Omega \to \mathbf{R}$ such that f is harmonic on $\text{Int } \Omega$ and $f = g$ on $\partial \Omega$.

We ask you to solve this problem by yourself. The following three exercises should be of some help.

1. Let $\mathcal{R}_{x_0}$ denote the set of all random walks on $\mathbf{Z}^2$ with x_0 as the starting point. This set can be identified with the set of all sequences of N's, E's, S's, and W's (for example, N W W E S N...). Assign to N, E, S, and W the numerical values $0, 1, 2,$ and 3. Let $I = (0, 1] = $ the half-closed unit interval. If $\omega \in I$, the quaternary expansion of ω gives rise to a sequence of 0's, 1's, 2's and 3's and hence to a sequence such as that above. Therefore, we can identify I with $\mathcal{R}_{x_0}$. (For the details of this identification, see §1.2.) Now suppose $x_0 \in \Omega$. Consider the random walk $r_\omega \in \mathcal{R}_{x_0}$ indexed by $\omega \in I$. Two possibilities exist: Either r_ω stays inside $\text{Int } \Omega$ forever, or it eventually gets to a boundary point $x_b(\omega)$. (For instance, if $x_0 \in \partial \Omega$, then $x_b(\omega) = x_0$.)
    a. Show that the first of these two possibilities occurs with probability zero. (See §1.4, exercise 17.)
    b. Let $f_{x_0}(\omega) = g[x_b(\omega)]$. Show that f_{x_0} is a measurable function of $\omega \in I$.

2. Let $\mathscr{R}^N_{x_0}$ be the set of all random walks starting at x_0 that move directly to x_N on the first step. Define $\mathscr{R}^E_{x_0}$, $\mathscr{R}^S_{x_0}$, and $\mathscr{R}^W_{x_0}$ similarly.

   a. Show that $\mathscr{R}_{x_0} = \mathscr{R}^N_{x_0} \cup \mathscr{R}^E_{x_0} \cup \mathscr{R}^S_{x_0} \cup \mathscr{R}^W_{x_0}$ (disjoint union) and show that, under the correspondence $\mathscr{R}_{x_0} \sim I$, $\mathscr{R}^N_{x_0}$ corresponds to the interval $(0, \frac{1}{4}]$, $\mathscr{R}^E_{x_0}$ to the interval $(\frac{1}{4}, \frac{1}{2}]$, and so on.

   b. There is an obvious bijective map $\rho : \mathscr{R}^N_{x_0} \to \mathscr{R}_{x_N}$. Namely, take the random walk whose first position after x_0 is x_N and think of it as a random walk starting at x_N. Show that, if we identify $\mathscr{R}_{x_N}$ with $(0, 1]$ as in exercise 1 and identify $\mathscr{R}^N_{x_0}$ with $(0, \frac{1}{4}]$ as in part a above, the mapping ρ becomes the mapping $\omega \to 4\omega$.

   c. Show that, with the identifications in parts a and b,

   $$(*) \qquad\qquad f_{x_N}(\omega) = f_{x_0}\!\left(\frac{\omega}{4}\right)$$

   Obtain comparable identities for f_{x_E}, f_{x_S}, and f_{x_W}.

3. Define $f : \Omega \to \mathbf{R}$ by setting

$$f(x_0) = \int_I f_{x_0}(\omega)\, d\mu$$

for all $x_0 \in \Omega$, with μ being Lebesgue measure. Prove that f is harmonic and equal to g on $\partial\Omega$.

Chapter 3

Fourier Analysis

§3.1 $\mathscr{L}^1$-Theory

Let $(X, \mathscr{F}, \mu)$ be a measure space and let $f : X \to \mathbf{C}$ be a complex-valued function. We can write $f(x) = u(x) + iv(x)$, where u and v are real-valued functions on X and $i = \sqrt{-1}$.

Definition 1. $f = u + iv$ is *measurable* if u and v are both measurable.

Note. If $f = u + iv$ is measurable, then $|f| = \sqrt{u^2 + v^2}$ is measurable by Theorem 14 in §2.1.

Proposition 2. Let $f = u + iv$ be measurable. The following two statements are equivalent.

1. $\displaystyle\int_X |f| \, d\mu < \infty$

2. $\displaystyle\int_X |u| \, d\mu < \infty \quad \text{and} \quad \int_X |v| \, d\mu < \infty$

Proof. Notice that $|u| + |v| \geq (u^2 + v^2)^{1/2} \geq |u|$ (or $|v|$). But $|f| = (u^2 + v^2)^{1/2}$, so integration yields

$$\int_X |u| \, d\mu + \int_X |v| \, d\mu \geq \int_X |f| \, d\mu$$

and

$$\int_X |f|\,d\mu \geq \int_X |u|\,d\mu \quad \left(\text{or} \int_X |v|\,d\mu \right)$$

The first of these inequalities shows that statement 2 implies statement 1; the second inequality shows that statement 1 implies statement 2. □

Definition 3. If $f = u + iv$ is a complex-valued measurable function on X, we say f is *integrable* if

$$\int_X |f|\,d\mu < \infty$$

In this case we define the *integral* of f to be the complex number

(1) $$\int_X f\,d\mu = \int_X u\,d\mu + i \int_X v\,d\mu$$

We denote by $\mathcal{L}^1(X, \mu)$ the set of all such functions:

(2) $$\mathcal{L}^1(X, \mu) = \left\{ f : X \to \mathbf{C}, \text{measurable;} \int_X |f|\,d\mu < \infty \right\}$$

Proposition 4. Let $f, g \in \mathcal{L}^1(X, \mu)$, $c \in \mathbf{C}$, then

1. $f + g \in \mathcal{L}^1(X, \mu)$ and $\displaystyle\int_X (f + g)\,d\mu = \int_X f\,d\mu + \int_X g\,d\mu$

2. $cf \in \mathcal{L}^1(X, \mu)$ and $\displaystyle\int_X (cf)\,d\mu = c \int_X f\,d\mu$

3. $\displaystyle\overline{\int_X f\,d\mu} = \int_X \bar{f}\,d\mu$

4. $\displaystyle\left| \int_X f\,d\mu \right| \leq \int_X |f|\,d\mu$

Proof. We leave proofs of 1, 2, and 3 as exercises. To prove 4 let $a = \int_X f\,d\mu$. Then $(\bar{a}/|a|) \int_X f\,d\mu = |a|$ is a positive real number. Let $g = (\bar{a}/|a|)f$ and write $g = u + iv$, where u and v are real valued. Then

$$\left| \int_X f\,d\mu \right| = |a| = \frac{\bar{a}}{|a|} \int_X f\,d\mu = \int_X g\,d\mu$$

$$= \int_X u\,d\mu + i \int_X v\,d\mu$$

Thus $\int_X v\,d\mu = 0$ because the left-hand side is real. Therefore

$$\left|\int_X f\,d\mu\right| = \int_X u\,d\mu \le \int_X |u|\,d\mu \le \int_X |g|\,d\mu \le \int_X |f|\,d\mu$$

since $|f| = |g|$.   □

Parts 1 and 2 of proposition 4 are a proof that the space $\mathscr{L}^1(X,\mu)$ is a vector space over the complex numbers. In fact, $\mathscr{L}^1(X,\mu)$ is a *normed* vector space.

Definition 5. Let V be a vector space over **C**. A *norm* on V is a function $\|\cdot\| : V \to \mathbf{R}$ with the following properties:

a. $\|v\| \ge 0,\ v \in V$
b. $\|v\| = 0 \Leftrightarrow v = 0$
c. $\|cv\| = |c|\,\|v\|,\ c \in \mathbf{C},\ v \in V$
d. $\|v + w\| \le \|v\| + \|w\|$

Given a norm $\|\cdot\|$ on a vector space V, we can define a metric $d(\cdot,\cdot):$ $V \times V \to \mathbf{R}$ on V by $d(v,w) = \|v - w\|$. It is easy to check that d satisfies the properties of a metric; that is,

1. $d(v,w) = d(w,v),\ v,w \in V$
2. $d(v,w) + d(w,u) \ge d(v,u),\ u,v,w \in V$
3. $d(v,w) = 0$ if and only if $v = w$

(See Appendix A for a review of metric spaces.)

There is a natural candidate for a norm on the space $\mathscr{L}^1(X,\mu)$.

Definition 6. Let $f \in \mathscr{L}^1(X,\mu)$. We define $\|f\|_1 = \int_X |f|\,d\mu$ to be the $\mathscr{L}^1$-norm of f.

Unfortunately, $\|\cdot\|$ does not quite satisfy property b. Instead it satisfies property b':

$$\|f\|_1 = 0 \Leftrightarrow f = 0 \text{ a.e.}$$

This statement means that two functions f and g in $\mathscr{L}^1(X,\mu)$ have to be considered the same if they are equal a.e. With this convention, it is easy to show the following theorem.

Theorem 7. $\|\cdot\|_1$ is a norm on $\mathscr{L}^1(X,\mu)$.

Proof. Properties a, b', and c are obvious. Property d follows by integrating the inequality

$$|f(x) + g(x)| \le |f(x)| + |g(x)|$$ □

We will say that a sequence of functions $f \in \mathcal{L}^1(X, \mu)$, $n = 1, 2, \ldots$, converges to a function $f \in \mathcal{L}^1(X, \mu)$ *in the $\mathcal{L}^1$-norm* (or simply *converges in
$\mathcal{L}^1$*) if

(3) $$\|f_n - f\|_1 \to 0 \quad \text{as} \quad n \to \infty$$

Convergence in this sense is *not* the same as pointwise convergence almost
everywhere. In the exercises, you will find examples of sequences that converge
in one of these two senses but fail to converge in the other. (See exercises 1
and 2.) The best one can conclude about the relationship between these two
notions of convergence is the following.

Theorem 8. Suppose f_n, $n = 1, 2, \ldots$, converges to f in the $\mathcal{L}^1$-norm.
Then there exists a subsequence f_{n_i}, $i = 1, 2, \ldots$, that converges to f almost
everywhere.

We will, in fact, prove a somewhat stronger result. Recall that, for a metric
space (V, d), a sequence $v_n \in V$, $n = 1, 2, \ldots$, is said to be a *Cauchy sequence* if
$d(v_m, v_n) \to 0$ as $m, n \to \infty$. In particular, a sequence of functions $f_n \in \mathcal{L}^1(X, \mu)$,
$n = 1, 2, \ldots$, is a Cauchy sequence if

(4) $$\|f_m - f_n\|_1 \to 0 \quad \text{as} \quad m, n \to \infty$$

Theorem 9. Let f_n, $n = 1, 2, \ldots$, be a Cauchy sequence in $\mathcal{L}^1$. Then there
exists a subsequence f_{n_i}, $i = 1, 2, \ldots$, that converges almost everywhere to an
$\mathcal{L}^1$ function f. In addition, the original sequence converges to f in the
$\mathcal{L}^1$-norm.

Proof. Choose n_1 such that, for $m, n > n_1$, $\|f_m - f_n\|_1 < \frac{1}{4}$. Next choose
$n_2 > n_1$ such that, for $m, n > n_2$, $\|f_m - f_n\|_1 < \frac{1}{8}$. Continuing inductively,
choose $n_i > n_{i-1}$ such that, for $m, n > n_i$, $\|f_m - f_n\|_1 < 1/2^{i+1}$. We will show
that the subsequence $\{f_{n_i}\}$ converges pointwise almost everywhere. By
construction

(5) $$\|f_{n_{i+1}} - f_{n_i}\|_1 < \frac{1}{2^{i+1}}$$

Let $g_1 = f_{n_1}$ and let $g_i = f_{n_i} - f_{n_{i-1}}$ for $i \geq 2$. Then

(6) $$f_{n_i} = \sum_{r=1}^{i} g_r$$

and $\|g_r\|_1 < 1/2^r$. Thus

$$\sum_{r=1}^{\infty} \int_X |g_r| \, d\mu < \infty$$

so, by the corollary of the Lebesgue dominated convergence theorem (corollary 12 of §2.3), the series

$$\sum_{r=1}^{\infty} g_r$$

converges pointwise almost everywhere; and, in view of equation 6, the sequence $\{f_{n_i}\}$ converges almost everywhere. Let f be the pointwise limit of the f_{n_i}'s. By assumption, f is defined for almost all $x \in X$, and we can define it at the remaining points of X by setting it equal to zero at these points. It remains for us to show that $f_n \to f$ in $\mathcal{L}^1$. Given $\varepsilon > 0$, there exists an n_0 such that

$$\int |f_m - f_n| \, d\mu < \varepsilon$$

for $m, n > n_0$. Fixing $n > n_0$ and letting $m \to \infty$, we get

$$\varepsilon \geq \liminf \int |f_m - f_n| \, d\mu \geq \int \liminf |f_m - f_n| \, d\mu$$

$$\geq \int |f - f_n| \, d\mu = \|f - f_n\|_1$$

by Fatou's lemma. Hence f_n converges to f in the $\mathcal{L}^1$-norm.   □

Recall that a metric space (V, d) is *complete* if every Cauchy sequence $v_n \in V$ has a limit $v \in V$. (Intuitively speaking, there are no "holes" in V.) A normed vector space $(V, \| \cdot \|)$ that is complete with respect to the metric

$$d(v, w) = \|v - w\|$$

is called a *Banach space*. By theorem 9, $\mathcal{L}^1(X, \mu)$ has this property, so we conclude the following.

Theorem 10. $\mathcal{L}^1(X, \mu)$ is a Banach space.

Exercises for §3.1

1. Let I be the unit interval $0 \leq x \leq 1$, and let $I_{k,n}$ be the subinterval

$$\frac{k}{n} \leq x \leq \frac{k+1}{n} \qquad 0 \leq k < n$$

Let f_1 be the characteristic function of $I_{0,1}$, f_2 and f_3 the characteristic functions of $I_{0,2}$ and $I_{1,2}$, f_3, f_4, and f_5 the characteristic functions of $I_{0,3}$, $I_{1,3}$, and $I_{2,3}$, and so on. Show that the sequence $\{f_n\}$ converges to 0 in $\mathcal{L}^1(I)$ but does not converge pointwise anywhere.

2. Let f_n be the function on the interval $(0, 1]$ that is equal to zero for $1/n \le x \le 1$ and is equal to n for $0 < x < 1/n$. Show that f_n converges pointwise to zero everywhere as $n \to \infty$ but does not converge in $\mathcal{L}^1$.

3. In exercise 1 extract a subsequence of the sequence $\{f_n\}$ that converges pointwise almost everywhere.

4. Let X be a finite interval and μ Lebesgue measure on X. Show that there exists a countable family of functions $\{f_i; i = 1, 2, 3, \ldots\}$ with the property that the f_i's are *dense* in $\mathcal{L}^1(X, \mu)$. That is, given any function $f \in \mathcal{L}^1(X, \mu)$ and any number $\varepsilon > 0$, then, for some f_i, $\|f_i - f\|_1 < \varepsilon$. (*Hint:* See §2.2, exercise 7.)

5. Let X be a set, and let $\mathscr{B}(X)$ be the set of all bounded, complex-valued functions on X. For $f \in \mathscr{B}(X)$ let

$$\|f\| = \sup_{x \in X} |f(x)|$$

Show that $\|\cdot\|$ is a norm, and show that $\mathscr{B}(X)$ is a Banach space with respect to this norm.

6. In exercise 5 suppose the set X is infinite. Show that, if $f_1, f_2, \ldots$ is a sequence of functions in $\mathscr{B}(X)$, there exists a function $f \in \mathscr{B}(X)$ such that

$$\|f - f_i\| \ge 1$$

for all i. (Compare with exercise 4.)

7. **a.** Let p and q be numbers greater than 1 with $(1/p) + (1/q) = 1$. Prove that

$$ab \le \frac{a^p}{p} + \frac{b^q}{q}$$

for any pair of nonnegative numbers a and b.

b. Let $(X, \mathscr{F}, \mu)$ be a measure space, and let f and g be nonnegative measurable functions. Prove that

(*) $$\int fg \, d\mu \le \left(\int f^p \, d\mu \right)^{1/p} \left(\int g^q \, d\mu \right)^{1/q}$$

(*Hint:* Let $\alpha = (\int f^p \, d\mu)^{1/p}$ and $\beta = (\int g^q \, d\mu)^{1/q}$. At each point $x \in X$, apply the inequality in (i) with $a = f(x)/\alpha$ and $b = g(x)/\beta$, and integrate with respect to x.)

8. Let f and g be as in exercise 7. Show that

$$\left(\int (f + g)^p \, d\mu \right)^{1/p} \le \left(\int f^p \, d\mu \right)^{1/p} + \left(\int g^p \, d\mu \right)^{1/p}$$

(*Hint:* Write $(f + g)^p = f(f + g)^{p-1} + g(f + g)^{p-1}$ and apply equation (*) to each of the two products $f(f + g)^{p-1}$ and $g(f + g)^{p-1}$.)

9. a. Let $1 \le p < \infty$. A complex-valued measurable function $f : X \to \mathbf{C}$ is said to be $\mathscr{L}^p$-integrable if $\int |f|^p \, d\mu < \infty$. Denote by $\mathscr{L}^p(X, \mu)$ the set of all such functions. Show that $\mathscr{L}^p(X, \mu)$ is a vector space. That is, show that, if f and g are in $\mathscr{L}^p(X, \mu)$, so is $f + g$, and that, if f is in $\mathscr{L}^p(X, \mu)$, any constant multiple of f is in $\mathscr{L}^p(X, \mu)$ as well.

b. If $f \in \mathscr{L}^p(X, \mu)$, let

$$|f|_p = \left(\int |f|^p \, d\mu \right)^{1/p}$$

Show that $\| \cdot \|_p$ is a norm on $\mathscr{L}^p(X, \mu)$.

§3.2 $\mathscr{L}^2$-Theory

Let $(X, \mathscr{F}, \mu)$ be a measure space. A measurable function $f : X \to \mathbf{C}$ is said to be $\mathscr{L}^2$-*integrable* or *square-integrable* if

$$(1) \qquad \int_X |f|^2 \, d\mu < \infty$$

We denote by $\mathscr{L}^2(X, \mu)$ the set of all such functions; that is,

$$(2) \qquad \mathscr{L}^2(X, \mu) = \left\{ f : X \to \mathbf{C}, \text{measurable}; \int_X |f|^2 \, d\mu < \infty \right\}$$

Definition 1. The quantity

$$(3) \qquad \| f \|_2 = \left(\int_X |f|^2 \, d\mu \right)^{1/2}$$

is called the $\mathscr{L}^2$-norm of $f \in \mathscr{L}^2(X, \mu)$.

We will see in a moment that equation 3 does indeed define a norm and that $\mathscr{L}^2$ is a Banach space with respect to this norm. First, however, we will establish a few elementary facts about $\mathscr{L}^2$.

Theorem 2. If f and g are in $\mathscr{L}^2(X, \mu)$, fg is in $\mathscr{L}^1(X, \mu)$.

Proof. Let $X_1 = \{x \in X; |f(x)| > |g(x)|\}$ and let $X_2 = \{x \in X; |g(x)| \ge |f(x)|\}$. Then, on X_1, $|fg| \le |f|^2$; and, on X_2, $|fg| \le |g|^2$. So

$$\int_X |fg| \, d\mu \le \int_{X_1} |f|^2 \, d\mu + \int_{X_2} |g|^2 \, d\mu$$

$$\le \| f \|_2^2 + \| g \|_2^2 \qquad \qquad \square$$

Corollary 3. If $\mu(X) < \infty$, then $\mathscr{L}^2(X, \mu)$ is contained in $\mathscr{L}^1(X, \mu)$.

Proof. If $\mu(X) < \infty$, the constant function 1 is in $\mathscr{L}^2(X, \mu)$.   □

Corollary 4. If f and g are in $\mathscr{L}^2(X, \mu)$, so is $f + g$.

Proof. It is enough to show that $|f + g|^2$ is in $\mathscr{L}^1$, but $|f + g|^2 \leq |f|^2 + 2|f||g| + |g|^2$.   □

Corollary 4 says that $\mathscr{L}^2(X, \mu)$ is a vector space over the complex numbers. We will soon show that it has some other nice properties as well. First, however, we need to discuss briefly the subject of *inner product spaces*.

Definition 5. A vector space V over the complex numbers is an *inner product space* if it is equipped with a mapping

$$\langle \cdot, \cdot \rangle : V \times V \to \mathbf{C}$$

such that

1. $\langle v_1 + v_2, w \rangle = \langle v_1, w \rangle + \langle v_2, w \rangle$
2. $\langle cv, w \rangle = c\langle v, w \rangle$
3. $\langle v, w \rangle = \overline{\langle w, v \rangle}$
4. $\langle v, v \rangle \geq 0$ and $\langle v, v \rangle = 0$ if and only if $v = 0$

An example of an inner product space with which you are already familiar is the finite dimensional space $\mathbf{C}^n$. If $v = (a_1, \ldots, a_n) \in \mathbf{C}^n$ and $w = (b_1, \ldots, b_n) \in \mathbf{C}^n$, the inner product of v and w is

$$(4) \qquad \sum_{i=1}^{n} a_i \overline{b_i}$$

We will show that the much more complicated space $\mathscr{L}^2(X, \mu)$ is also an inner product space. Indeed, by theorem 2, the quantity

$$(5) \qquad \langle f, g \rangle = \int_X f\overline{g} \, d\mu$$

is well-defined for $f, g \in \mathscr{L}^2(X, \mu)$; and it is obvious from proposition 4 of §3.1 that it satisfies properties 1–3. It doesn't quite satisfy property 4; in fact, if

$$\langle f, f \rangle = \int |f|^2 \, d\mu = 0$$

the most we can conclude is that $f = 0$ a.e. But, if we put in force the convention that an $\mathscr{L}^2$ function is zero "in the $\mathscr{L}^2$ sense" when it is zero a.e., then property 4 is reinstated and we have proved the following theorem.

Theorem 6. $\mathscr{L}^2(X, \mu)$, equipped with the inner product given by equation 5, is an inner product space.

We now prove a few facts that are true for inner product spaces in general and, thus, for $\mathscr{L}^2(X, \mu)$ in particular. Given an inner product space $(V, \langle \ , \ \rangle)$ and $v \in V$, let

$$(6) \qquad\qquad\qquad \|v\| = \sqrt{\langle v, v \rangle}$$

By property 4 this is well-defined and is zero if and only if $v = 0$. We call $\|v\|$ the *norm* of the vector $v \in V$.

Theorem 7.   (Schwarz's inequality) If $v, w \in V$, then

$$(7) \qquad\qquad\qquad |\langle v, w \rangle| \le \|v\| \, \|w\|$$

Proof.   If $w = 0$, the inequality is obvious. If $w \ne 0$, consider, for $\lambda \in \mathbf{C}$,

$$0 \le \langle v + \lambda w, v + \lambda w \rangle = \langle v, v \rangle + \bar{\lambda} \langle v, w \rangle + \lambda \overline{\langle v, w \rangle} + |\lambda|^2 \langle w, w \rangle$$

Letting $\lambda = -\langle v, w \rangle / \langle w, w \rangle$, this inequality implies

$$0 \le \langle v, v \rangle - \frac{|\langle v, w \rangle|^2}{\langle w, w \rangle}$$

or, equivalently

$$|\langle v, w \rangle|^2 \le \langle v, v \rangle \langle w, w \rangle \qquad\qquad \square$$

Corollary 8.   (Triangle inequality) For $v, w \in V$

$$(8) \qquad\qquad\qquad \|v + w\| \le \|v\| + \|w\|$$

Proof.   Squaring equation 8 we get

$$(9) \qquad\qquad\qquad \|v + w\|^2 \le \|v\|^2 + 2\|v\| \, \|w\| + \|w\|^2$$

But $\|v + w\|^2 = \langle v + w, v + w \rangle = \langle v, v \rangle + \langle v, w \rangle + \langle w, v \rangle + \langle w, w \rangle = \|v\|^2 + 2Re\langle v, w \rangle + \|w\|^2$, so equation 9 reduces to the inequality

$$2Re\langle v, w \rangle \le 2\|v\| \, \|w\|$$

which is an immediate consequence of inequality 7. $\square$

From this corollary we conclude the following.

Corollary 9.   The norm $\|\cdot\|$ on V is a norm in the sense of definition 5 of §3.1; that is, $(V, \|\cdot\|)$ is a normed vector space.

In particular, we restate this result for the vector space $\mathscr{L}^2(X, \mu)$.

Corollary 10. $\mathcal{L}^2(X, \mu)$ is a normed vector space.

Moreover, applying Schwarz's inequality to $\mathcal{L}^2(X, \mu)$, we deduce the following.

Corollary 11. If f and g are in $\mathcal{L}^2$, then

(10) $$\| fg \|_1 \leq \| f \|_2 \| g \|_2$$

An inner product space $(V, \langle \ , \ \rangle)$ that is complete with respect to the norm $\| \cdot \|$ (that is, one that is a Banach space with respect to this norm) is called a *Hilbert space*. For example, $\mathbf{C}^n$ is a Hilbert space.

We will show that $\mathcal{L}^2(X, \mu)$ is a Hilbert space. To simplify the proof we will make an assumption about the underlying measure space $(X, \mathcal{F}, \mu)$. We recall from the last paragraph of §1.3 the following definition.

Definition 12. A measure space $(X, \mathcal{F}, \mu)$ is σ-finite if there exists a sequence $X_n \in \mathcal{F}, n = 1, 2, \dots$ with $\bigcup_{n=1}^{\infty} X_n = X$ and $\mu(X_n) < \infty$.

Without loss of generality, one can assume that $X_1 \subseteq X_2 \subseteq \cdots$.

Theorem 13. $\mathcal{L}^2(X, \mu)$ is complete with respect to the norm $\| \cdot \|_2$; that is, it is a Hilbert space.

Proof. Assume that X is σ-finite. (See exercise 9 for a way to get rid of this assumption.) Let $\{ f_n \}$ be a Cauchy sequence in $\mathcal{L}^2(X, \mu)$. Choose X_n's as in definition 12. By corollary 3, $\{ f_n \}$ is a Cauchy sequence in $\mathcal{L}^1(X_1, \mu)$. So, by theorem 9 of §3.1, we can extract from $\{ f_n \}$ a subsequence $\{ f_{1,n} \}$ that converges a.e. on X_1. Repeating the process, extract from this subsequence a smaller subsequence $\{ f_{2,n} \}$ that converges a.e. on X_2. Continuing inductively, one obtains for each i a subsequence $\{ f_{i,n} \}$ of $\{ f_{i-1,n} \}$ that converges a.e. on X_i. Now apply the Cantor diagonal process: The subsequence $f_{1,1}, f_{2,2}, \dots$ converges a.e. on X and its pointwise limit is equal a.e. to a measurable function g. Let $g_1 = f_{1,1}$, $g_2 = f_{2,2}$, and so on. Because the g_n's are a subsequence of the f_n's, they are also a Cauchy sequence in $\mathcal{L}^2(X, \mu)$. So, for any $\varepsilon > 0$, there exists an n_0 such that $\| g_m - g_n \|_2^2 < \varepsilon$ when $m, n > n_0$.

By Fatou's lemma, with $n > n_0$ fixed and $m \to \infty$,

$$\int \liminf |g_m - g_n|^2 \, d\mu \leq \liminf \int |g_m - g_n|^2 \, d\mu \leq \varepsilon$$

But the term on the left is

$$\int |g - g_n|^2 \, d\mu$$

because g_m converges to g pointwise a.e. Hence, we conclude that g is in

$\mathscr{L}^2(X, \mu)$ and that g_n converges to g in $\mathscr{L}^2(X, \mu)$. Finally, if a subsequence of a Cauchy sequence converges, the sequence itself converges as well, so f_n also converges to g. □

The following general fact about inner product spaces will be useful in §3.3.

Theorem 14. Let V be an inner product space. Then the inner product $\langle \cdot, \cdot \rangle$ is continuous in both variables with respect to the norm given by equation 6. In other words, if $v_n \to v$ and $w_n \to w$, then $\langle v_n, w_n \rangle \to \langle v, w \rangle$.

Proof. If $v_n \to v$ with respect to $\| \cdot \|$, then $\| v_n - v \| \leq 1$ for n large, so

$$\| v_n \| \leq \| v_n - v \| + \| v \| \leq 1 + \| v \|$$

for n large. Then

$$|\langle v_n, w_n \rangle - \langle v, w \rangle| \leq |\langle v_n, w_n \rangle - \langle v_n, w \rangle| + |\langle v_n, w \rangle - \langle v, w \rangle|$$

$$\leq |\langle v_n, w_n - w \rangle| + |\langle v_n - v, w \rangle|$$

$$\leq \| v_n \| \| w_n - w \| + \| v_n - v \| \| w \|$$

$$\leq (1 + \| v \|) \| w_n - w \| + \| v_n - v \| \| w \|$$

Hence $|\langle v_n, w_n \rangle - \langle v, w \rangle|$ tends to zero as $n \to \infty$. □

Exercises for §3.2

1. Let V be an inner product space. Show that Schwarz's inequality

$$|\langle v, w \rangle| \leq \| v \| \| w \|$$

is an equality if and only if $w = 0$ or $v = cw$ for some complex number c.

2. Let V be an an inner product space. Show that

$$(*) \qquad \| v + w \|^2 + \| v - w \|^2 = 2(\| v \|^2 + \| w \|^2)$$

(Geometrically, the sum of the squared lengths of the diagonals in the figure below is equal to the sum of the squared lengths of the sides.)

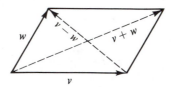

3. Let V be a normed vector space whose norm satisfies the identity in equation (∗) in exercise 2. Show that there exists an inner product on V such that

$$\|v\| = \sqrt{\langle v, v \rangle}$$

(*Hint:* Show that

$$2\mathrm{Re}\langle v, w \rangle = \|v + w\|^2 - \|v\|^2 - \|w\|^2$$

if an inner product exists.)

4. Let $X = (0, 1]$, equipped with Lebesgue measure. Show that the function $f(x) = x^{-3/4}$ is in $\mathscr{L}^1(X, \mu)$ but not in $\mathscr{L}^2(X, \mu)$.

5. Let $X = [1, \infty)$, equipped with Lebesgue measure. Show that $f(x) = x^{-3/4}$ is in $\mathscr{L}^2(X, \mu)$ but not in $\mathscr{L}^1(X, \mu)$.

6. Let $a_1, a_2, \ldots$ be a sequence of positive numbers with $\sum_{n=1}^{\infty} a_n = \infty$. Let $S_n = \sum_{i=1}^n a_i$. Show that

$$\sum_{n=1}^{\infty} \frac{a_n}{S_n} = \infty$$

but that

$$\sum_{n=1}^{\infty} \frac{a_n}{S_n^2} < \infty$$

(*Hint:* Let $f_n = \sum_{i=1}^n a_i/s_i$ and let $g_n = \sum_{i=1}^n a_i/s_i^2$. Compare f_n with $\log s_n$ and g_n with $1/s_n$.)

7. Let $(X, \mathscr{F}, \mu)$ be a measure space that is σ-finite but not finite. Show that $\mathscr{L}^2(X, \mu)$ is not contained in $\mathscr{L}^1(X, \mu)$. (*Hint:* Use exercise 6.)

8. Let $(X, \mathscr{F}, \mu)$ be a measure space and $f_1, f_2, \ldots$ a sequence of $\mathscr{L}^2$ functions on X. Let

$$X' = \{x \in X; f_i(x) \neq 0 \text{ for some } i\}$$

Show that X' is σ-finite; that is, show that it is a countable union of measurable sets of finite measure. (*Hint:* Let $E_{m,n} = \{x \in X; |f_n(x)| > 1/m\}$. Show that $X' = \bigcup_{m,n} E_{m,n}$.)

9. Using exercise 8, show that theorem 13 is still true without the hypothesis that X is σ-finite.

10. Let $(X, \mathscr{F}, \mu)$ be a measure space. Prove that $\mathscr{L}^p(X, \mu)$ is a Banach space— that is, complete with respect to the norm $\|\cdot\|_p$. (See §3.1, exercise 9.)

11. (Sobolev's inequality) Let f be a function on the interval $[0, 1]$ that is continuous and has a continuous first derivative f'. Show that

$$\sup_{0 \leq x, y \leq 1} |f(x) - f(y)| \leq \|f'\|_2$$

§3.3 The Geometry of Hilbert Space

In this section we will discuss some of the geometric properties of a Hilbert space $\mathscr{L}$ with inner product $\langle\ ,\ \rangle$. When we apply this material in the later sections of this chapter, $\mathscr{L}$ will always be $\mathscr{L}^2(X,\mu)$ where $(X,\mathscr{F},\mu)$ is a measure space.

Definition 1.   If $f,g\in\mathscr{L}$, we say f is *orthogonal* to g (written $f\perp g$) if $\langle f,g\rangle = 0$.

Theorem 2.   (Pythagoras) If f, $g\in\mathscr{L}$ with $f\perp g$, then $\|f\|^2 + \|g\|^2 = \|f+g\|^2$.

Proof.

$$\|f+g\|^2 = \langle f+g, f+g\rangle$$

$$= \langle f,f\rangle + \langle f,g\rangle + \langle g,f\rangle + \langle g,g\rangle$$

$$= \|f\|^2 + \|g\|^2 \qquad\qquad\qquad\square$$

More generally, suppose that $f_1,f_2,\ldots,f_n\in\mathscr{L}$ with $f_i\perp f_j$, $i\neq j$. Then by induction it is easy to prove that $\|f_1 + f_2 + \cdots + f_n\|^2 = \sum_{i=1}^n \|f_i\|^2$.

One basic example of a Hilbert space you should always keep in mind is $\mathbf{C}^n$ with the inner product given by equation 4 of §3.2. This Hilbert space has finite "dimension." We will see, however, that some Hilbert spaces are "infinite dimensional"; in fact, these are the spaces that are most interesting to us.

Definition 3.   A sequence $\phi_1,\phi_2,\phi_3,\ldots$ in $\mathscr{L}$ is called *orthonormal* if

(1)
$$\langle\phi_i,\phi_j\rangle = \begin{cases}1 & \text{if } i = j \\ 0 & \text{if } i \neq j\end{cases}$$

Example 4.   Let $\mathscr{L}$ be $\mathbf{C}^n$ with the inner product given by equation 4 of §3.2. Let $v_1 = (1,0,\ldots,0)$, $v_2 = (0,1,0,\ldots,0),\ldots, v_n = (0,\ldots,0,1)$. Then $v_1,\ldots,v_n$ is an orthonormal sequence.

Example 5.   Let $X = [-\pi,\pi]$, $\mu = $ Lebesgue measure, and $\mathscr{L} = \mathscr{L}^2(X,\mu)$. Let $\phi_k = (1/\sqrt{2\pi})e^{ikx}$, $-\infty < k < \infty$. It is easy to check that the ϕ_k's form an orthonormal sequence. Indeed

$$\int_X \phi_k\overline{\phi}_j\,d\mu = \frac{1}{2\pi}\int_{-\pi}^{\pi} e^{i(k-j)x}\,d\mu$$

$$= \begin{cases}1 & \text{if } k = j \\ 0 & \text{if } k \neq j\end{cases}$$

Recall now that an *orthonormal basis* in $\mathbf{C}^n$ is, by definition, an orthonormal sequence of vectors $v_1, v_2, \ldots, v_n$. This definition depends on the dimension of $\mathbf{C}^n$. Another characterization of orthonormal basis in $\mathbf{C}^n$ allows us to turn the tables and determine the dimension from the length of the basis: Namely, if $v_1, v_2, \ldots, v_n$ is a basis of $\mathbf{C}^n$, then there are no nonzero vectors that are simultaneously orthogonal to all the v_i's. This motivates the following definition.

Definition 6. An orthonormal sequence $\phi_1, \phi_2, \ldots$ is called *complete* if, for any $f \in \mathscr{L}$, the conditions

$$f \perp \phi_i \qquad i = 1, 2, \ldots$$

imply $f = 0$.

Remark. If $\mathscr{L} = \mathscr{L}^2(X, \mu)$, we must interpret $f = 0$ as $f = 0$ a.e.

Definition 7. Let $\mathscr{L}$ be a Hilbert space, and suppose that $\phi_1, \phi_2, \ldots, \phi_n$ is a complete orthonormal sequence in $\mathscr{L}$. Then $\mathscr{L}$ is said to have *dimension n*. If $\mathscr{L}$ contains an infinite orthonormal sequence $\phi_1, \phi_2, \ldots$, $\mathscr{L}$ is said to be *infinite dimensional*.

We leave it to the reader to check that the dimension of a Hilbert space is well-defined (see exercise 1).

Remark. Example 5 shows that, if $X = [-\pi, \pi]$ and $\mu = \mu_L$, then $\mathscr{L}^2(X, \mu)$ is infinite dimensional. We will see in §3.4 that the ϕ_k's described in example 5 are complete.

In infinite dimensions the notion of completeness of an orthonormal sequence replaces the notion of an orthonormal basis for finite dimensions. We now study some of the properties of a complete orthonormal sequence in $\mathscr{L}$.

Let $\phi_1, \phi_2, \ldots$ be a complete orthonormal sequence in $\mathscr{L}$. Given $f \in \mathscr{L}$, let $c_i = \langle f, \phi_i \rangle$; this is called the ith *Fourier coefficient* of f with respect to the sequence $\phi_1, \phi_2, \ldots$. The formal series $\sum_{i=1}^{\infty} c_i \phi_i$ is called the *Fourier series* of f with respect to the sequence $\phi_1, \phi_2 \ldots$.

Theorem 8. Let $\phi_1, \phi_2, \ldots$ be a complete orthonormal sequence in $\mathscr{L}$. Let $f \in \mathscr{L}$ and let $c_i = \langle f, \phi_i \rangle$. Define $S_n = \sum_{i=1}^{n} c_i \phi_i$ to be the nth partial sum of the Fourier series of f. Then $S_n \to f$ in $\mathscr{L}$.

Proof. Write $f = f - S_n + S_n$. Notice that $\langle f - S_n, \phi_i \rangle = 0$ as long as $i \leq n$ because

$$\langle f, \phi_i \rangle = c_i = \langle S_n, \phi_i \rangle \qquad \text{for } i \leq n$$

Thus $(f - S_n) \perp S_n$ because S_n is a linear combination of ϕ_i's with $i \leq n$. Then, by theorem 2,

$$\|f\|^2 = \|f - S_n\|^2 + \|S_n\|^2$$

$$\geq \|S_n\|^2$$

Now, $$\|S_n\|^2 = \langle S_n, S_n \rangle = \left\langle \sum_{i=1}^{n} c_i\phi_i, \sum_{i=1}^{n} c_i\phi_i \right\rangle$$

$$= \sum_{i=1}^{n} |c_i|^2$$

So, we have shown that

(2) $$\sum_{i=1}^{n} |c_i|^2 \leq \|f\|^2 \qquad \text{for all } n$$

That is, $\sum_{i=1}^{\infty} |c_i|^2$ converges.

Now consider $S_n - S_m$ for $n > m$. $S_n - S_m = \sum_{i=m+1}^{n} c_i\phi_i$, so by theorem 2

$$\|S_n - S_m\|^2 = \sum_{i=m+1}^{n} |c_i|^2$$

Because we have shown that $\sum_{i=1}^{\infty} |c_i|^2$ converges, we can conclude that the sequence of S_n's is Cauchy. Because $\mathscr{L}$ is complete, we know that there is a $g \in \mathscr{L}$ such that $S_n \to g$ in $\mathscr{L}$.

To finish the proof we need to show that $f = g$. Notice that, by theorem 14 of §3.2, $\langle g, \phi_i \rangle = \lim_{n \to \infty} \langle S_n, \phi_i \rangle = c_i$. Thus $\langle f - g, \phi_i \rangle = c_i - c_i = 0$ for all i. Because the ϕ_i's are complete, we conclude that $f = g$. □

Theorem 9. (Plancherel) Let $\phi_1, \phi_2, \ldots$ be a complete orthonormal sequence in $\mathscr{L}$. For $f \in \mathscr{L}$, let $c_i = \langle f, \phi_i \rangle$. Then $\|f\|^2 = \sum_{i=1}^{\infty} |c_i|^2$.

Proof. $S_n \to f$ in $\mathscr{L}$ so, by the continuity of the inner product,

$$\langle S_n, S_n \rangle \to \langle f, f \rangle = \|f\|^2$$

But $\langle S_n, S_n \rangle = \sum_{i=1}^{n} |c_i|^2$, so we conclude that

$$\sum_{i=1}^{\infty} |c_i|^2 = \|f\|^2 \qquad\qquad □$$

You may have noticed that the proof of theorem 8 did not use the completeness of the orthonormal sequence until the final line. When the sequence is not necessarily complete, we get the following.

Theorem 10. Let $\phi_1, \phi_2, \ldots$ be an orthonormal sequence in $\mathscr{L}$. For $f \in \mathscr{L}$, let $c_i = \langle f, \phi_i \rangle$ and $S_n = \sum_{i=1}^{n} c_i \phi_i$. Then S_n is Cauchy in $\mathscr{L}$ and thus converges to a limit $g \in \mathscr{L}$. Moreover, $\langle g, \phi_i \rangle = \langle f, \phi_i \rangle$ for all i.

Proof. This proof is the same as the proof of theorem 8 with the last sentence omitted. $\quad\square$

If the sequence $\{\phi_n\}$ is not complete, we get the following in place of the Plancherel theorem.

Theorem 11. (Bessel's inequality) Let f and c_i be as in theorem 10. Then

$$\|f\|^2 \geq \sum_{i=1}^{\infty} |c_i|^2$$

Proof. This inequality is a direct consequence of equation 2. $\quad\square$

Corollary 12. The Fourier coefficients c_i tend to zero as $i \to \infty$.

An important example of a Hilbert space is the space whose elements are infinite sequences

$$(3) \qquad s = (a_1, a_2, a_3, \ldots)$$

of complex numbers satisfying

$$(4) \qquad \sum_{i=1}^{\infty} |a_i|^2 < \infty$$

The set of all such sequences is denoted l^2 (read as "little $\mathscr{L}^2$"). It is easy to see that it is a Hilbert space. In fact, it is a Hilbert space of the form $\mathscr{L}^2(X, \mu)$. Take for X the set of positive integers—that is, $X = \{1, 2, 3, \ldots\}$. Let $\mathscr{F}$ be the σ-field of all subsets of X, and let μ be the counting measure:

$$\mu(A) = \text{number of points in } A$$

A function on X is just a sequence such as in equation 3. In the exercises we will ask you to show that such a sequence is in $\mathscr{L}^2(X, \mu)$ if and only if equation 4 holds. We will also ask you to show that for two such sequences

$$s = (a_1, a_2, a_3, \ldots)$$

and

$$t = (b_1, b_2, b_3, \ldots)$$

their inner product is

(5)
$$\sum_{i=1}^{\infty} a_i \bar{b}_i$$

(Compare with equation 4 of §3.2.) Notice that Schwarz's inequality for l^2 says that

(6)
$$\left| \sum_{i=1}^{\infty} a_i \bar{b}_i \right| \le \left(\sum_{i=1}^{\infty} |a_i|^2 \right)^{1/2} \left(\sum_{i=1}^{\infty} |b_i|^2 \right)^{1/2}$$

We will use this fact in §3.4.

Notice that, if we are given *any* Hilbert space $\mathscr{L}$ and a complete orthonormal sequence $\phi_1, \phi_2, \ldots$ in $\mathscr{L}$, then, by the Plancherel formula, the sequence of Fourier coefficients

$$s = (c_1, c_2, \ldots)$$

is in l^2. Conversely, we claim that, given a sequence

$$(c_1, c_2, c_3, \ldots) \in l^2$$

the sequence of partial sums

$$S_n = \sum_{i=1}^{n} c_i \phi_i$$

converges in $\mathscr{L}$ to a limiting element f. Indeed, for $m > n$

$$\| S_m - S_n \|^2 = \sum_{i=n+1}^{m} |c_i|^2$$

so S_n is a Cauchy sequence and the assertion follows from the completeness of $\mathscr{L}$. Thus, if we have a complete (infinite) orthonormal sequence in $\mathscr{L}$, *we get a bijective map of $\mathscr{L}$ onto l^2.* This statement is rather surprising in view of the fact that $\mathscr{L}$ can be, in principle, a much more complicated space than l^2—for example, the space of square-integrable functions on $\mathbf{R}^n$.

Exercises for §3.3

1. Show that the dimension of a Hilbert space is well-defined.
2. Let X be the set of positive integers and μ the counting measure on X. Show that $\mathscr{L}^2(X, \mu) = l^2$. Moreover, show that the $\mathscr{L}^2$ inner product on $\mathscr{L}^2(X, \mu)$ is the inner product given by equation 5.
3. Let f be an $\mathscr{L}^2$ function on the interval $[-\pi, \pi]$. Show that

$$\int_{-\pi}^{\pi} f(x) e^{-inx}\, dx \to 0 \quad \text{as} \quad n \to \infty$$

(*Hint:* See corollary 12.)

4. Let $(X, \mathscr{F}, \mu)$ be a measure space. A sequence $f_n \in \mathscr{L}^2(X, \mu), n = 1, 2, \ldots,$ is said to converge in the $\mathscr{L}^2$ sense to $f \in \mathscr{L}^2(X, \mu)$ if $\|f - f_n\|_2 \to 0$ as $n \to \infty$. Prove that, if $f_n \to f$ in the $\mathscr{L}^2$ sense, there exists a subsequence f_{n_i}, $i = 1, 2, 3, \ldots,$ that converges to f a.e.

5. Let X be the unit interval and μ Lebesgue measure. Show that $\mathscr{L}^2$-convergence in $\mathscr{L}^2(X, \mu)$ does *not* imply pointwise convergence a.e., and vice versa.

6. ($\mathscr{L}^2$-convergence of the randomized harmonic series) Let μ be Lebesgue measure on the unit interval, and let R_n be the nth Rademacher function. Let

$$S_n = \sum_{i=1}^{n} \left(\frac{1}{i}\right) R_i$$

   Show that S_n converges in the $\mathscr{L}^2$ sense to a function $H \in \mathscr{L}^2(I, \mu)$.

7. For fixed m and n with $m > n$, let A be the set

   $$\{\omega \in I; |S_k(\omega) - S_n(\omega)| > \varepsilon \text{ for some } k \text{ between } m \text{ and } n\}$$

   Prove that

   $$\mu(A) \le \frac{1}{\varepsilon^2} \int (S_m - S_n)^2 \, d\mu$$

   Here are some hints:

   (i) With k fixed, let $J \subset I$ be a union of intervals of the form $i/2^k < t \le (i + 1)/2^k$, with i between zero and $2^k - 1$. Show that, if $n \le k < m$, $\int_J R_m R_n \, d\mu = 0$.

   (ii) If J is as in part i, show that

   $$\int_J (S_m - S_k)(S_k - S_n) \, d\mu = 0$$

   and also show that

   $$\int_J (S_m - S_n)^2 \, d\mu \ge \int_J (S_k - S_n)^2 \, d\mu$$

   (iii) For $n \le k \le m$, let A_k be the set

   $$\{\omega \in I; |S_j(\omega) - S_n(\omega)| \le \varepsilon \text{ for } n \le j < k \text{ and } |S_k(\omega) - S_n(\omega)| > \varepsilon\}$$

   Show that $A = \bigcup_{k=n}^{m} A_k$ (disjoint union) and

   $$\mu(A_k) \le \frac{1}{\varepsilon^2} \int_{A_k} (S_m - S_n)^2 \, d\mu$$

8. Using exercise 7 prove that the randomized harmonic series

$$\sum_{i=1}^{\infty} \left(\frac{1}{i}\right) R_i(\omega)$$

converges almost everywhere to the function H of exercise 6.

9. (The Gram–Schmidt process)

a. Let $\mathscr{L}$ be an inner product space and $f_1, f_2, \ldots, f_n$ elements of $\mathscr{L}$. Show that, if $f_1, \ldots, f_n$ are linearly independent, then there exists an orthonormal sequence $\phi_1, \ldots, \phi_n$ such that, for all i,

(*) $\phi_i = \sum_{j \le i} a_{ij} f_i$

with $a_{ii} > 0$.

(*Hint:* Let $\phi_1 = c_1 f_1$ where c_1^{-1} is the length of f_1. Let $\phi_2 = c_2[f_2 - (f_2, \phi_1)\phi_1]$ where c_2^{-1} is the length of $f_2 - (f_2, \phi_1)\phi_1$. Continue.)

b. Let $f_1, f_2, \ldots$ be an infinite sequence of elements of $\mathscr{L}$. Suppose that, for all n, $f_1, \ldots, f_n$ are linearly independent. Show that there exists an orthonormal sequence $\phi_1, \phi_2, \ldots$ such that equation (*) holds for all i.

10. a. Let $\mathscr{L} = \mathscr{L}^2([-1, 1])$. Let $f_1 = 1, f_2 = x, f_3 = x^2$, and so on. Apply the Gram–Schmidt process to this sequence (see exercise 9). Show that the resulting ϕ_i's are polynomial functions of x. (These functions are called the *Legendre polynomials.*) Compute the first few of these functions.

Answer: $\phi_1 = \dfrac{1}{\sqrt{2}}$ $\phi_2 = \sqrt{\dfrac{3}{2}} x$

$\phi_3 = \sqrt{\dfrac{5}{2}} \left(\dfrac{3}{2} x^2 - 1\right)$ $\phi_4 = \sqrt{\dfrac{7}{2}} \left(\dfrac{5}{2} x^3 - \dfrac{3}{2} x\right)$

b. Use the Weierstrass approximation theorem (see §2.7, exercise 9) to show that the Legendre polynomials are a complete orthonormal sequence in $\mathscr{L}^2([-1, 1])$.

11. a. Show that the *Haar* functions,

$$H_{0,0}(x) = 1 \qquad \text{for } 0 \le x \le 1$$

and

$$H_{n,k}(x) = \begin{cases} -2^{n/2} & \text{for } \dfrac{k-1}{2^n} \le x < \dfrac{k-\frac{1}{2}}{2^n} \\[2ex] 2^{n/2} & \text{for } \dfrac{k-\frac{1}{2}}{2^n} \le x < \dfrac{k}{2^n} \\[2ex] 0 & \text{elsewhere} \end{cases}$$

where $n \geq 1$ and $1 \leq k \leq 2^n$, are an orthonormal sequence in $\mathcal{L}^2([0, 1])$.

b. Suppose that $f \in \mathcal{L}^2([0,1])$ and $(f, H_{n,k}) = 0$ for all n, k. Show that, if A is an interval of the form $[k/2^n, l/2^n]$, with $0 \leq k < l \leq 2^n$, then

(**)
$$\int_A f \, d\mu = 0$$

c. Show that equation (**) holds for every subinterval A of $[0, 1]$. Conclude that $\{H_{n,k}\}$ is a complete orthonormal sequence.

d. Let R_{n+1} be the $n + 1$st Rademacher function. Show that

$$R_{n+1} = \frac{1}{2^n} \sum_{k=1}^{2^n} H_{n,k}$$

§3.4 Fourier Series

We pointed out in §3.3 that the functions

$$\phi_n = \left(\frac{1}{\sqrt{2\pi}} \right) e^{inx} \qquad -\infty < n < \infty$$

form an orthonormal sequence in the space $\mathcal{L}^2(-\pi, \pi)$. We will show in this section that this orthonormal sequence is complete. In the course of proving this result, we will also prove a number of classical results about convergence of Fourier series. To begin, note that, by corollary 12 of §3.3,

(1)
$$c_n = \langle f, \phi_n \rangle = \frac{1}{\sqrt{2\pi}} \int_{-\pi}^{\pi} f(x) e^{-inx} \, dx$$

tends to zero as $|n| \to \infty$, provided that f is an $\mathcal{L}^2$ function on the interval $[-\pi, \pi]$.

Let f be a measurable function defined on the whole real line. We will say that f is periodic of period 2π if

(2)
$$f(x + 2\pi) = f(x) \quad \text{a.e.}$$

Given any measurable function defined on the interval $(-\pi, \pi]$, one can extend it uniquely to a periodic function on the whole real line by requiring that equation 2 hold. Moreover, if f is periodic of period 2π and integrable over the interval $[-\pi, \pi]$, then by equation 2 it is integrable over every compact subinterval of the real line. In fact, if I is a subinterval of length 2π, then

(3)
$$\int_I f \, dx = \int_{-\pi}^{\pi} f \, dx$$

To see this, suppose that I is of the form $[a - \pi, a + \pi]$. Then

$$\int_I f(x)\,dx = \int_\pi^{\pi+a} f(x)\,dx + \int_{-\pi}^{\pi} f(x)\,dx - \int_{-\pi}^{-\pi+a} f(x)\,dx$$

But by periodicity the first term and third term cancel.

We will first study convergence of Fourier series for functions that are rather nicely behaved. Let f be a continuous function periodic of period 2π, and let x_0 be a point on the interval $[-\pi, \pi]$. Suppose that the right and left derivatives of f exist at x_0; that is, the limits

(4)$_+$
$$\lim_{x \to (x_0)_+} \frac{f(x) - f(x_0)}{x - x_0}$$

and

(4)$_-$
$$\lim_{x \to (x_0)_-} \frac{f(x) - f(x_0)}{x - x_0}$$

exist. We will prove the following theorem.

Theorem 1. The series

$$\frac{1}{\sqrt{2\pi}} \sum_{n=-\infty}^{\infty} c_n e^{inx}$$

converges at $x = x_0$ and its limit is $f(x_0)$.

Proof. Let $S_N(f)(x_0)$ be the Nth partial sum of this series. By equation 1

$$S_N(f)(x_0) = \frac{1}{\sqrt{2\pi}} \sum_{k=-N}^{N} c_k e^{ikx_0}$$

$$= \frac{1}{2\pi} \int_{-\pi}^{\pi} f(x) \left(\sum_{k=-N}^{N} e^{ik(x_0 - x)} \right) dx$$

Setting

(5)
$$D_N(x) = \frac{1}{2\pi} \sum_{k=-N}^{N} e^{ikx}$$

we get for $S_N(f)(x_0)$ the formula

$$S_N(f)(x_0) = \int_{-\pi}^{\pi} f(x) D_N(x_0 - x)\,dx$$

Making the change of coordinates $x_0 - x \to x$, this integral becomes

$$\int_{x_0-\pi}^{x_0+\pi} f(x_0 - x)D_N(x)\,dx$$

So, by equation 3 we get finally

(6) $$S_N(f)(x_0) = \int_{-\pi}^{\pi} f(x_0 - x)D_N(x)\,dx$$

To estimate the right-hand side of this formula, we note first the following properties of $D_N(x)$:

(7) $$\int_{-\pi}^{\pi} D_N(x)\,dx = 1$$

and

(8) $$D_N(x) = \frac{1}{2\pi}\frac{e^{i(N+1)x} - e^{-iNx}}{e^{ix} - 1}$$

Proof of properties 7 and 8. To obtain equation 7, just integrate equation 5 term by term and note that all terms except $k = 0$ have integral zero. To obtain equation 8, rewrite $D_N(x)$ as

$$\frac{1}{2\pi}e^{-iNx}\left(\sum_{k=0}^{2N} e^{ikx}\right)$$

and note that, with $\alpha = e^{ix}$, the second factor is just

$$\sum_{k=0}^{2N} (\alpha)^k \qquad \text{or} \qquad \frac{\alpha^{2N+1} - 1}{\alpha - 1}$$

Next we note that the denominator in equation 8 has a zero of first order at $x = 0$. In fact,

$$\lim_{x\to 0}\frac{e^{ix} - 1}{x} = \frac{d}{dx}(e^{ix})|_{x=0} = i$$

and $e^{ix} - 1$ has no zeroes on the interval $[-\pi, \pi]$ except at $x = 0$; so the function

$$\frac{x}{e^{ix} - 1}$$

is continuous on this interval providing we define it to be $-i$ at $x = 0$.

We now return to the proof of theorem 1. By equation 7

$$f(x_0) = \int_{-\pi}^{\pi} f(x_0) D_N(x) \, dx$$

Subtracting this from equation 6, we get

(9) $$S_N(f)(x_0) - f(x_0) = \int_{-\pi}^{\pi} [f(x_0 - x) - f(x_0)] D_N(x) \, dx$$

Set

$$g(x) = \frac{f(x_0 - x) - f(x_0)}{e^{ix} - 1} = \frac{f(x_0 - x) - f(x_0)}{x} \left(\frac{x}{e^{ix} - 1} \right)$$

The second factor on the right is continuous on the interval $[-\pi, \pi]$, as we just observed. The first factor is continuous except at $x = 0$, and at $x = 0$ it is continuous on the left and on the right by assumptions $4_\pm$. Hence g is piecewise continuous and *a fortiori* in $\mathcal{L}^2$. By equations 8 and 9

$$S_N(f)(x_0) - f(x_0) = \frac{1}{2\pi} \int_{-\pi}^{\pi} g(x) e^{i(N+1)x} \, dx - \frac{1}{2\pi} \int_{-\pi}^{\pi} g(x) e^{-iNx} \, dx$$

The first term on the right is the $-(N + 1)$st Fourier coefficient of g, and the second term is the Nth Fourier coefficient; so by equation 1 both these terms tend to zero as $N \to \infty$. □

A function f is called *piecewise differentiable* if its domain of definition is a finite union of closed intervals and if, on each of these intervals, df/dx exists and is continuous. It is clear that, if f is piecewise differentiable, it satisfies assumptions $4_\pm$ at all points in its domain of definition. We will show that for such functions theorem 1 can be considerably improved.

Theorem 2.   Let f be a continuous function that is periodic of period 2π. If f is piecewise differentiable on the interval $[-\pi, \pi]$, then $S_N(f)$ converges to f uniformly and absolutely on this interval.

Proof.   Let g be the derivative of f. By assumption, g is defined and continuous on the interval $[-\pi, \pi]$ except at a finite number of points, and we will define it everywhere by defining it arbitrarily at these points. We will first show that, if $c_n(f)$ and $c_n(g)$ are the nth Fourier coefficients of f and g, respectively, then

(10) $$c_n(g) = inc_n(f)$$

Proof of equation 10.   We can find $a_0 = -\pi < a_1 < a_2 < \cdots < a_r = \pi$ such that g is continuous on $[a_i, a_{i+1}]$. Then

$$\int_{-\pi}^{\pi} g e^{-inx}\, dx = \sum \int_{a_i}^{a_{i+1}} g e^{-inx}\, dx = \sum \int_{a_i}^{a_{i+1}} \left(\frac{df}{dx}\right) e^{-inx}\, dx$$

$$= \sum \left[f e^{-inx} \Big|_{a_i}^{a_{i+1}} + in \int_{a_i}^{a_{i+1}} f e^{-inx}\, dx \right]$$

$$= f e^{-inx} \Big|_{-\pi}^{\pi} + in \int_{-\pi}^{\pi} f e^{-inx}\, dx$$

However, the first term vanishes because f is periodic.

Remark. Integration by parts is justified on $[a_i, a_{i+1}]$ because this integral equals the Riemann integral.

To prove theorem 2 it is enough to show that $\sum |c_n(f)| < \infty$. Indeed, because we already know by theorem 1 that $S_n(f) \to f$ pointwise, this fact will imply that the convergence is absolute and uniform. Because g is in $\mathscr{L}^2$, $\sum |c_n(g)|^2 < \infty$; so by equation 10

$$\sum n^2 |c_n(f)|^2 < \infty$$

Therefore, by Schwarz's inequality for l^2 (see equation 6 of §3.3),

$$\sum_{n \neq 0} |c_n(f)| = \sum_{n \neq 0} \left(\frac{1}{|n|}\right) |n| |c_n| \leq \left(\sum_{n \neq 0} \frac{1}{n^2}\right)\left(\sum n^2 |c_n|^2\right)$$

Because both terms on the right are finite, so is their product. □

Let's now return to the proof that the functions $\phi_n = (1/\sqrt{2\pi})e^{inx}$ form a complete orthonormal sequence. We have to show that, if f is an $\mathscr{L}^2$ function with $\langle f, \phi_n \rangle = 0$ for all n, then $f = 0$ a.e. We will first show that, if f has this property, then

(11)
$$\int_a^b f\, dx = 0$$

for every subinterval $[a, b]$ of $[-\pi, \pi]$. Let $\varepsilon > 0$ and let χ_ε be the function indicated in the following figure:

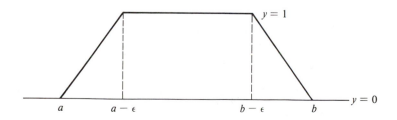

This function is piecewise differentiable. So, by theorem 2, $S_N(\chi_\varepsilon) \to \chi_\varepsilon$ uniformly and, hence, *a fortiori* in $\mathscr{L}^2$. Thus

$$0 = \langle f, S_N(\chi_\varepsilon) \rangle \to \langle f, \chi_\varepsilon \rangle$$

that is, $\langle f, \chi_\varepsilon \rangle = 0$. As ε tends to zero, χ_ε converges in $\mathscr{L}^2$ to the characteristic function of the interval $[a, b]$. So, by a repetition of the preceding argument,

$$\lim_{\varepsilon \to 0} \int_{-\pi}^{\pi} f \chi_\varepsilon \, dx = \int_a^b f \, dx = 0$$

Thus, we have established equation 11.

Consider now the collection of all measurable subsets of the interval $[-\pi, \pi]$ for which

$$(12) \qquad \int_A f \, dx = 0$$

This collection contains all the subintervals of $[-\pi, \pi]$ and is a λ-system; so, by the $\pi-\lambda$ theorem (theorem 7 of §2.5), it contains all Borel subsets of $[-\pi, \pi]$. Because every measurable set is a disjoint union of a Borel set and a set of measure zero, equation 12 holds for *all* measurable sets A. In particular, let A_+ be the set where $f > 0$ and A_- be the set where $f \le 0$. Then

$$\int_{-\pi}^{\pi} |f| \, dx = \int_{A_+} f \, dx - \int_{A_-} f \, dx = 0$$

so $f = 0$ a.e.

The Plancherel theorem now gives us

$$(13) \qquad \int_{-\pi}^{\pi} |f|^2 \, dx = \sum_{n=-\infty}^{\infty} |c_n|^2$$

with

$$(14) \qquad c_n = \frac{1}{\sqrt{2\pi}} \int_{-\pi}^{\pi} f(x) e^{-inx} \, dx$$

for $f \in \mathscr{L}^2(-\pi, \pi)$. We will discuss some applications of this identity in the exercises.

Exercises for §3.4

1. a. Let f be an integrable function on the interval $[-\pi, \pi]$. Let

$$f_M(x) = \begin{cases} f(x) & \text{when } |f(x)| \le M \\ 0 & \text{when } |f(x)| > M \end{cases}$$

Show that $\int |f_M - f| \, d\mu \to 0$ as $M \to \infty$.

b. Show that the nth Fourier coefficient of f tends to zero as $n \to \infty$. (*Hint:* What can you say about the nth Fourier coefficient of f_M?)

2. Let f be a periodic function of period 2π possessing continuous derivatives up to order k. Show that

$$c_n\left(\frac{d^k f}{dx^k}\right) = (in)^k c_n(f)$$

3. **a.** Let f be the function

$$f(x) = \begin{cases} 0 & \text{for } -\pi \le x < 0 \\ 1 & \text{for } 0 \le x \le \pi \end{cases}$$

What are its Fourier coefficients?

b. Prove that

$$\sum_{n=1}^{\infty} \frac{1}{(2n-1)^2} = \frac{\pi^2}{8} \quad \text{and} \quad \sum_{n=1}^{\infty} \frac{1}{n^2} = \frac{\pi^2}{6}$$

4. The zeta function

$$\zeta(s) = \sum_{n=1}^{\infty} n^{-s} \qquad s > 1$$

is of considerable importance in number theory. By judicious use of the Plancherel theorem, evaluate this function at $s = 2$ and $s = 4$. Can you devise a method for evaluating $\zeta(s)$ at *all* even integers? (*Hint:* See exercise 2.)

5. **a.** Let $f = f(x, t)$ be a function that has continuous second derivatives in x and t and is periodic of period 2π in x. Let $c_n(t)$ be the nth Fourier coefficient of $f(x, t)$, regarded as a function of x (that is, with t fixed). Show that, if f is a solution of the heat equation

(*)
$$\frac{\partial f}{\partial t} = \frac{\partial^2 f}{\partial x^2}$$

then $c_n(t) = e^{-n^2 t} c_n(0)$.

b. Given a function $f_0(x)$ that is periodic of period 2π and has continuous second derivative, show how to construct a solution of the heat equation (*) with initial data: $f(x, 0) = f_0(x)$.

6. The Weierstrass approximation theorem says that, if f is a continuous function on the interval $[a, b]$ and $\varepsilon > 0$, there exists a polynomial p with

$$\sup_{a \le x \le b} |f(x) - p(x)| < \varepsilon$$

We sketched a proof of this theorem in exercises 8 and 9 of §2.7. Deduce from theorem 2 a second proof. Here are some hints:

(i) Show that one can assume $-\pi < a < b < \pi$.

(ii) Show that every continuous function f on the interval $[a, b]$ can be extended to a continuous function on $[-\pi, \pi]$ that is periodic of period 2π.

(iii) Show that, if f is a continuous function that is periodic of period 2π, there exists a piecewise differentiable function f_0 that is periodic of period 2π and is $\varepsilon/4$ close to f; that is,

$$\sup_{-\pi \le x \le \pi} |f - f_0| < \frac{\varepsilon}{4}$$

(*Hint:* See figure.)

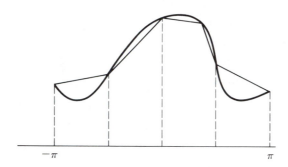

(iv) Let $S_n(x) = \sum_{-N}^{N} c_n e^{inx}$, the Nth partial sum of the Fourier series for f_0. Show that, for N sufficiently large,

$$\sup_{-\pi \le x \le \pi} |S_N - f| < \frac{\varepsilon}{2}$$

(v) In the formula for S_N, replace e^{inx} by $\sum_{r=0}^{k} (1/r!)(inx)^r$ with k large.

7. Show that in theorem 1 it is enough to assume that f is continuous and differentiable from the left and from the right at x_0 and is *piecewise* continuous elsewhere.

8. a. Compute the Fourier series of the sawtooth function

$$s(x) = \begin{cases} x - \pi & \text{for } 0 < x \le \pi \\ x + \pi & \text{for } -\pi < x \le 0 \end{cases}$$

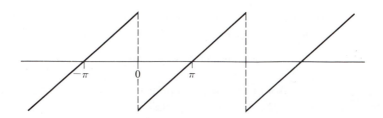

b. Show that the series

$$\sum_{n\neq 0} \frac{1}{n} e^{inx}$$

converges everywhere on the interval $-\pi \leq x \leq \pi$ except at the origin.

9. Let f and g be $\mathscr{L}^2$ functions on the interval $(-\pi, \pi]$. Extend them to functions on **R** by requiring that they be periodic of period 2π. Show that the *convolution*

$$(f * g)(x) = \frac{1}{2\pi} \int_{-\pi}^{\pi} f(y)g(x - y)\,dy$$

is in $\mathscr{L}^1(-\pi, \pi)$ and that its Fourier coefficients c_n are just

$$c_n = a_n b_n$$

where a_n and b_n are the Fourier coefficients of f and g.

§3.5 The Fourier Integral

Let f be a complex-valued integrable function defined on the real line. Its Fourier transform is the function

(1) $$\hat{f}(y) = \int f(x)e^{-ixy}\,dx$$

Notice that this function is well-defined because the absolute value of the integrand is $|f(x)|$. Indeed

$$|\hat{f}(y)| = \left| \int f(x)e^{-ixy}\,dx \right| \leq \int |f(x)|\,dx$$

so $\hat{f}(y)$ is bounded by the $\mathscr{L}^1$-norm of f. (In exercise 2 you will be asked to show that $\hat{f}(y)$ is continuous and that $\hat{f}(y) \to 0$ as $y \to \pm\infty$.)

On the interval $(-\pi, \pi)$, $\mathscr{L}^2$-integrable functions are automatically $\mathscr{L}^1$-integrable; however, for functions defined on the real line, this is no longer the case (see §3.2, exercise 5). Therefore, equation 1 does not make sense if the integrand is an arbitrary $\mathscr{L}^2$ function. Nevertheless, we will show that equation 1 can be appropriately defined for $\mathscr{L}^2$ functions and that, just as for Fourier series, the $\mathscr{L}^2$-theory of the Fourier integral is remarkably simple and elegant.

We will start by studying the Fourier transform for a very well-behaved class of functions.

Definition 1. Let f be a complex-valued function defined on the real line whose derivatives of all orders—that is, df/dx, d^2f/dx^2, and so on—exist and are continuous. Then f is called a *Schwartz function* if, for each pair of nonnegative integers m and n, there exists a constant C (depending on m and n) such that

(2)
$$\left| x^m \left(\frac{d^n f}{dx^n} \right) \right| \le C$$

We will denote by S the set of all Schwartz functions. If f and g are in S, so is $f + g$; and, if f is in S, constant multiples of f are in S and so are xf and df/dx. Also, equation 2 implies that, given N, there is a constant C, depending on N, so that

(3)
$$|f(x)| \le C(1 + |x|^2)^{-N}$$

So Schwartz functions go to zero very rapidly as $x \to \pm \infty$. The basic example of a Schwartz function, about which we will have much to say in the next two sections, is the function $e^{-(x^2/2)}$.

By equation 3 Schwartz functions are $\mathscr{L}^1$-integrable; so their Fourier transforms are defined. We will prove that *the Fourier transform of a Schwartz function is again a Schwartz function*. To see this fact we need a fundamental property of the Fourier transform.

Lemma 2.

1. Let $f \in S$ and let $g(x) = xf(x)$. Then $\hat{g}(y) = \sqrt{-1}(d/dy)\hat{f}(y)$.
2. Let $f \in S$ and let $h = df/dx$. Then $\hat{h}(y) = \sqrt{-1}\,y\hat{f}(y)$.

In other words, up to factors of $\sqrt{-1}$, the Fourier transform interchanges the operations "differentiation by x" and "multiplication by x."

Proof. By definition

$$\hat{f}(y) = \int f(x)e^{-ixy}\,dx$$

The integrand on the right is differentiable with respect to y, and the derivative is again integrable; so the left side is differentiable with respect to y, and

$$\frac{d\hat{f}(y)}{dy} = \int \left(\frac{d}{dy} \right) [f(x)e^{-ixy}]\,dx$$

$$= -i \int xf(x)e^{-ixy}\,dx$$

$$= -i\hat{g}(y)$$

which proves part 1. To prove part 2 we note that

$$\hat{h}(y) = \int \left(\frac{d}{dx}\right) f(x) e^{-ixy} \, dx$$

$$= -\int f(x) \left(\frac{d}{dx}\right) e^{-ixy} \, dx$$

$$= iy \int f(x) e^{-ixy} \, dx$$

The integration by parts is justified by the fact that f is going to zero very rapidly as $x \to \pm\infty$. ▽

It follows from the lemma that, if f is a Schwartz function, then $(d/dy)\hat{f}$ and $y\hat{f}$ are the Fourier transforms of Schwartz functions, and by induction $y^m(d^n/dy^n)\hat{f}$ is the Fourier transform of a Schwartz function for all m and n. In particular, $y^m(d^n/dy^n)\hat{f}$ is bounded; so $\hat{f}$ is a Schwartz function.

Example. Let $f = e^{-(x^2/2)}$. We will show that

(4) $$\hat{f}(y) = \sqrt{2\pi} \, e^{-(y^2/2)}$$

That is, up to a constant, f is its own Fourier transform.

Proof. Notice that f satisfies the differential equation

(5) $$\frac{df}{dx} + xf = 0$$

Indeed, up to a constant factor, f is the *only* solution of this equation, for, if

$$\frac{dh}{dx} + xh = 0$$

then $(d/dx)e^{x^2/2}h = e^{x^2/2}[(dh/dx) + xh] = 0$. So $e^{x^2/2}h$ is equal to a constant C and $h = Ce^{-(x^2/2)}$. By lemma 2, $\hat{f}$ satisfies equation 5 if f does; so $\hat{f}(y)$ is a constant multiple of $e^{-(y^2/2)}$. All that remains to check is that this constant is $\sqrt{2\pi}$. But, if $\hat{f}(y) = Ce^{-(y^2/2)}$, then

$$C = \hat{f}(0) = \int e^{-(y^2/2)} \, dy$$

The integral on the right can be evaluated by elementary means and shown to be $\sqrt{2\pi}$. ☐

We can now state the first main result of this section.

Theorem 3. The mapping $f \to \hat{f}$ maps S bijectively onto itself. Moreover, if $f \in S$ and $g = \hat{f}$, then $f = \check{g}$ where

(6)
$$\check{g}(x) = \frac{1}{2\pi} \int g(y) e^{ixy} \, dy$$

Remarks.

1. The function $\check{g}$ in display 6 is called the *inverse Fourier transform* of g. Notice that it is very simply related to the Fourier transform of g—namely,

(7)
$$\check{g}(y) = \left(\frac{1}{2\pi}\right) \hat{g}(-y)$$

From this identity it is clear that the inverse Fourier transform maps the Schwartz space into itself.

2. Equation 6 implies that the Fourier transform is *injective* as a map of S into S. (That is, if $f \in S$ and $g = \hat{f} = 0$, then, by equation 6, $f = 0$.) It also implies that the inverse Fourier transform is *surjective* as a map of S into S. (That is, if $f \in S$ and $g = \hat{f}$, then $f = \check{g}$.) But, because of the simple relation (equation 7) between the Fourier transform and its inverse, we conclude that both the Fourier transform and the inverse Fourier transform are injective and surjective. So, if we can prove equation 6, we will have automatically proved the rest of theorem 1. Incidentally, equation 6 is usually referred to as the *Fourier inversion formula*.

For the proof of equation 6 we will need some additional properties of the Fourier transform.

Lemma 4. Let f and g be Schwartz functions. Then

(8)
$$\int \hat{f}(y) g(y) \, dy = \int f(x) \hat{g}(x) \, dx$$

Proof. By Fubini's theorem

$$\int \hat{f}(y) g(y) \, dy = \int \left(\int f(x) e^{-ixy} \, dx \right) g(y) \, dy$$

$$= \int \left(\int g(y) e^{-ixy} \, dy \right) f(x) \, dx$$

$$= \int \hat{g}(x) f(x) \, dx$$

The interchange of integrations is justified by equation 3. $\triangledown$

Lemma 5. Let $f \in S$ and let a be a real number. Then, if $f_a(x) = f(x + a)$,

(9) $$\hat{f}_a(y) = e^{iay} \hat{f}(y)$$

Proof. By definition

$$\hat{f}_a(y) = \int f(x + a) e^{-ixy} \, dx$$

So, if we make the change of variables $x = s - a$, this becomes

$$\hat{f}_a(y) = e^{iay} \int f(s) e^{-isy} \, ds = e^{iay} \hat{f}(y) \qquad \triangledown$$

Lemma 6. Let $f \in S$ and let a be a positive number. Then, if $f_a(x) = f(x/a)$,

(10) $$\hat{f}_a(y) = a\hat{f}(ay)$$

Proof. By definition

$$\hat{f}_a(y) = \int f\left(\frac{x}{a}\right) e^{-ixy} \, dx$$

So, if we make the change of variables $x = as$, this becomes

$$\hat{f}_a(y) = a \int f(s) e^{-iasy} \, ds = a\hat{f}(ay) \qquad \triangledown$$

We will now prove equation 6. Let $f = f(x)$ be an arbitrary Schwartz function, and let $g = e^{-(y^2/2a^2)}$. Then, combining equations 4, 8, and 10, we get

(11) $$\int \hat{f}(y) e^{-(y^2/2a^2)} \, dy = \sqrt{2\pi}\, a \int f(x) e^{-(a^2 x^2/2)} \, dx$$

When we make the substitution $ax = s$, the right side becomes

(12) $$\sqrt{2\pi} \int f\left(\frac{s}{a}\right) e^{-s^2/2} \, ds$$

Now let $a \to +\infty$. Then $e^{-(y^2/2a^2)}$ tends to $e^0 = 1$ uniformly on compact sets; so the left side of equation 11 tends to $\int \hat{f}(y)\, dy$. On the other hand, by display 12, the right side of equation 11 tends to

$$\sqrt{2\pi}\, f(0) \int e^{-(s^2/2)} \, ds = 2\pi f(0)$$

and we obtain

(13) $$f(0) = \frac{1}{2\pi} \int \hat{f}(y) \, dy$$

Now let g be the function $g(x) = f(x + a)$. Replacing f by g in equation 13 and taking into account equation 9 we get

$$f(a) = g(0) = \frac{1}{2\pi} \int \hat{g}(y) \, dy = \frac{1}{2\pi} \int \hat{f}(y) e^{iay} \, dy \qquad \square$$

Next we will show that the Fourier transform preserves $\mathscr{L}^2$-norms up to a scalar factor.

Theorem 7. Let f be in S. Then

(14) $$\| \hat{f} \|_2^2 = 2\pi \| f \|_2^2$$

Proof. If we take complex conjugates of both sides of the identity

$$f(x) = \frac{1}{2\pi} \int e^{ixy} \hat{f}(y) \, dy$$

we get

$$\overline{f(x)} = \frac{1}{2\pi} \int e^{-ixy} \overline{[\hat{f}(y)]} \, dy$$

That is, $2\pi \overline{f}$ is the Fourier transform of $\overline{\hat{f}}$. Let $g = \overline{\hat{f}}$, so $\hat{g} = 2\pi \overline{f}$. Then, by lemma 2,

$$\int |\hat{f}|^2 \, dy = \int \hat{f} \overline{\hat{f}} \, dy = \int \hat{f} g \, dy = \int f \hat{g} \, dx$$

$$= \int f(2\pi \overline{f}) \, dx = 2\pi \int |f|^2 \, dx \qquad \square$$

Remark. This identity is called the *Plancherel formula* for the Fourier transform.

Using theorems 3 and 7 we can now define the Fourier transform of an arbitrary $\mathscr{L}^2$ function. The idea for this definition is based on a theorem about metric spaces: Suppose M and N are metric spaces and A is a dense subset of M. A map $f : A \to N$ is called *uniformly continuous* if, for every $\varepsilon > 0$, there exists a $\delta > 0$ such that $d_N(f(x), f(y)) < \varepsilon$ whenever $d_M(x, y) < \varepsilon$. The theorem we will need is the following.

Proposition 8. If $f : A \to N$ is uniformly continuous and N is complete, there exists a unique continuous mapping $g : M \to N$ extending f.

A proof of this fact is outlined in Appendix A.

By the Plancherel theorem, $\|\hat{f} - \hat{g}\|_2^2 = 2\pi\|f - g\|_2^2$; so the Fourier transform is uniformly continuous as a map of S into $\mathscr{L}^2$. Moreover, as we saw in §3.2, $\mathscr{L}^2$ is complete. Therefore, if we can show that S is dense in $\mathscr{L}^2$, then, by proposition 8, there is a unique extension of the Fourier transform from S to $\mathscr{L}^2$. In other words, if we can show that S is dense in $\mathscr{L}^2$, we will have succeeded in our goal of extending the Fourier transform to $\mathscr{L}^2$ functions.

To show that S is dense in $\mathscr{L}^2$, we need to show that there is a large supply of Schwartz functions. The results that we describe next make this point. We say that a function $f: \mathbf{R} \to \mathbf{R}$ is C^∞ if all of its derivatives—that is, df/dx, d^2f/dx^2, and so on—exist and are continuous.

Lemma 9. There exists a C^∞ function f_0 that is zero for $x \leq 0$ and positive for $x > 0$.

Proof. The function

$$f_0(x) = \begin{cases} e^{-(1/x)} & \text{for } x > 0 \\ 0 & \text{for } x \leq 0 \end{cases}$$

has this property.   ▽

Lemma 10. Given an interval (a, b), there exists a C^∞ function f_1 such that $f_1 = 0$ for $x \notin (a, b)$ and $f_1 > 0$ for $x \in (a, b)$.

Proof. Let f_0 be as in lemma 9 and let

$$f_1(x) = f_0(x - a)f_0(b - x)$$ ▽

Lemma 11. Given an interval (a, b), there exists a C^∞ function f_2 such that $f_2(x) = 0$ for $x \leq a$, $f_2(x) = 1$ for $x \geq b$, and $0 < f_2 < 1$ on the interval (a, b).

Proof. Let

$$f_2(x) = \frac{\displaystyle\int_{-\infty}^{x} f_1(s)\,ds}{\displaystyle\int_{-\infty}^{\infty} f_1(s)\,ds}$$ ▽

Lemma 12. Given $\varepsilon > 0$, there exists a C^∞ function f of the type depicted in the following figure.

Proof. By lemma 11 there exists a C^∞ function g such that $g = 0$ for $x \le a$, $g = 1$ for $x \ge a + \varepsilon$, and $0 < g < 1$ on the interval $(a, a + \varepsilon)$. Similarly, there exists a C^∞ function h such that $h = 0$ for $x \le b - \varepsilon$, $h = 1$ for $x \ge b$ and $0 < h < 1$ on the interval $(b - \varepsilon, b)$. Now let $f(x) = g(x)[1 - h(x)]$. ▽

Let $\bar{S}$ be the closure of S in $\mathscr{L}^2$; that is, $f \in \bar{S}$ if and only if there exists a sequence $f_n \in S$, $n = 1, 2, \ldots$, such that $\| f_n - f \|_2 \to 0$. It is clear that, if g and h are in $\bar{S}$, the $g + h$ is in $\bar{S}$ and constant multiples of g are in $\bar{S}$.

Proposition 13. Let A be a finite union of intervals. Then the characteristic function 1_A of A is in $\bar{S}$.

Proof. It is enough to prove this proposition for an interval $A = [a, b]$. Given $\varepsilon > 0$, let f be as in lemma 12. Because f is C^∞ and is zero outside $[a, b]$, f is in S. Moreover, $|f - 1_A| < 1$ on $(a, a + \varepsilon)$ and $(b - \varepsilon, b)$, and $f = 1_A$ elsewhere, so

$$\int |f - 1_A|^2 \, dx < 2\varepsilon \qquad \square$$

Proposition 14. Let A be a measurable set of finite measure. Then the characteristic function 1_A of A is in $\bar{S}$.

Proof. Choose $\varepsilon > 0$. By the definition of $\mathscr{M}_F$ (see §1.3), there exists a finite union of intervals B such that $\mu(S(A, B)) < \varepsilon$. Then

$$\int |1_B - 1_A|^2 \, dx = \mu(S(A, B)) < \varepsilon \qquad \square$$

Now let f be a nonnegative $\mathscr{L}^2$ function. By theorem 6 of §2.2, there exists an increasing sequence of simple functions $s_n \ge 0$ such that $s_n \to f$. By proposition 14, $s_n \in \bar{S}$. Moreover

$$\int |f - s_n|^2 \, dx \to 0 \quad \text{as} \quad n \to \infty$$

by the monotone convergence theorem, so $f \in \bar{S}$. Finally, let f be a complex-valued $\mathscr{L}^2$ function. Then

$$f = Re(f)_+ - Re(f)_- + \sqrt{-1}\,Im(f)_+ - \sqrt{-1}\,Im(f)_-$$

so f is in $\bar{S}$. Thus we have proved the following theorem.

Theorem 15. S is dense in $\mathscr{L}^2$.

We can now prove the second main result of this section.

Theorem 16. There is a unique linear mapping

(15) $$\hat{} : \mathscr{L}^2 \to \mathscr{L}^2$$

and a unique linear mapping

(16) $$\check{} : \mathscr{L}^2 \to \mathscr{L}^2$$

such that, restricted to S, equations 15 and 16 are the usual Fourier transform and inverse Fourier transform. These mappings are bijective and satisfy the Fourier inversion formula

$$f = (\hat{f})^{\check{}}$$

and the Plancherel formula

$$\|\hat{f}\|_2^2 = 2\pi \|f\|_2^2$$

Proof. We have already indicated how the Fourier transform can be extended to $\mathscr{L}^2$. The inverse Fourier transform can be extended the same way. Moreover, the Fourier inversion formula and the Plancherel formula hold on S; so, by continuity, they hold on $\mathscr{L}^2$. $\qquad\square$

Remark. For a general $\mathscr{L}^2$ function f, the integral in equation 1 doesn't make sense. How then do you evaluate $\hat{f}$? Of course, you have to approximate f by functions for which equation 1 does make sense, and then take limits. See exercise 4 for an explicit way to carry out this manipulation.

Exercises for §3.5

1. Show that, if $f \in \mathscr{L}^1(\mathbf{R})$, then for every $\varepsilon > 0$ there exists a Schwartz function g such that $\|f - g\|_1 < \varepsilon$.
2. If f is in $\mathscr{L}^1(\mathbf{R})$, show that $\hat{f}$ is continuous and $\hat{f}(\xi) \to 0$ as $\xi \to \pm\infty$. (*Hint:* These assertions are true when $f \in S$. Now use exercise 1.)
3. Show that, if f is both in $\mathscr{L}^1(\mathbf{R})$ and in $\mathscr{L}^2(\mathbf{R})$, the two definitions of $\hat{f}$—that is, equation 1 and the definition by continuity—coincide.

4. a. For $f \in \mathcal{L}^2(\mathbf{R})$ and $M > 0$, let

$$f_M(x) = \begin{cases} f(x) & \text{when } |x| \le M \\ 0 & \text{when } |x| > M \end{cases}$$

Show that $\| f_M - f \|_2 \to 0$ as $M \to \infty$.

b. Show that, if $f \in \mathcal{L}^2(\mathbf{R})$, then

(*) $\displaystyle \lim_{M \to \infty} \int_{-M}^{M} f(x) e^{-ixy} \, dx$

exists, in the sense of $\mathcal{L}^2$, and is equal to $\hat{f}$. (Equation (*) is often used as the definition of the $\mathcal{L}^2$ Fourier transform.)

5. Show that, if $f, g \in \mathcal{L}^2(\mathbf{R})$, then

$$\langle f, g \rangle = \left(\frac{1}{2\pi} \right) \langle \hat{f}, \hat{g} \rangle$$

(This identity is called *Parseval's identity*.) (*Hint:* The real part of $\langle f, g \rangle$ is equal to $\frac{1}{2}(\| f + g \|_2^2 - \| f \|_2^2 - \| g \|_2^2)$.)

6. a. Compute the Fourier transform of $xe^{-x^2/2}$ and of $x^2 e^{-x^2/2}$. Can you devise a scheme for computing the Fourier transform of $x^m e^{-x^2/2}$ for any m?

b. Show that for every integer m there exists a polynomial $H_m(x)$ of order m such that the Fourier transform of $H_m(x) e^{-x^2/2}$ is a constant multiple of itself. Moreover, show that one can choose the H_m's so that the sequence $H_m e^{-x^2/2}$, $m = 1, 2, \ldots$, is orthonormal. (The polynomial $H_m(x)$ is called the mth *Hermite polynomial*.) (*Hint:* Use exercise 5.)

7. a. Let c be a positive number and let f_c be the function

$$f_c(x) = \begin{cases} e^{-cx} & \text{for } x \ge 0 \\ 0 & \text{for } x < 0 \end{cases}$$

Show that its Fourier transform is $1/(c + iy)$.

b. Use the Plancherel formula to compute the integral

$$\int_{-\infty}^{\infty} \frac{dy}{c^2 + y^2}$$

8. a. Let f be the characteristic function of the interval $[-1, 1]$. Show that its Fourier transform is $(2 \sin y)/y$.

b. Use the Plancherel formula to compute

$$\int_{-\infty}^{\infty} \left(\frac{\sin x}{x} \right)^2 dx$$

9. a. If f and g are in $\mathcal{L}^1$, the *convolution* of f and g is the function

$$(f * g)(x) = \int f(x - y)g(y) \, dy$$

Show that $f * g$ is in $\mathscr{L}^1(\mathbf{R})$ and that $\| f * g \|_1 \leq \| f \|_1 \| g \|_1$. (*Hint:* Use the Fubini theorem.)

b. Show that the Fourier transform of $f * g$ is the product $\hat{f}\hat{g}$.

c. Conclude from part b that the convolution operation is associative and commutative.

10. a. Show that the function

$$g(x, t) = \left(\frac{1}{\sqrt{4\pi t}} \right) e^{-x^2/4t}$$

satisfies the heat equation $\partial g/\partial t = \partial^2 g/\partial x^2$ for $0 < t < \infty$ and $-\infty < x < \infty$.

b. Show that, if $f \in S$, the function

$$\mu(x, t) = g_t * f = \int g(t, x - y)f(y) \, dy$$

satisfies the heat equation for $0 < t < \infty$ and $-\infty < x < \infty$.

c. Show that, as $t \to 0+$, $\mu(x, t) \to f(x)$. (*Hint:* Using part b of the previous exercise, show that $\int |\hat{\mu}_t(y) - \hat{f}(y)| \, dy \to 0$ as $t \to 0+$. Here $\mu_t(x) = \mu(x, t)$.)

11. a. Let X be a set, $\mathscr{F}$ be a σ-field of subsets of X, and μ a probability measure on X. Given a random variable $f : X \to \mathbf{R}$, the function

$$\chi_f(t) = \int_X e^{itf} \, d\mu$$

is called the *characteristic function* of f. Show that χ_f is continuous and $|\chi_f(t)| \leq 1$.

b. Suppose f is bounded. Show that all the derivatives—$(d/dt)\chi_f$, $(d^2/dt^2)\chi_f$, and so on—exist and are continuous. Show that

$$\left(\frac{1}{i} \right)^n \left(\frac{d^n \chi_f}{dt^n} \right)(0) = \int_X f^n \, d\mu$$

12. a. Let $(X, \mathscr{F}, \mu)$ be as in exercise 11. Show that, if two random variables f and g are identically distributed, then $\chi_f = \chi_g$.

b. Conversely, show that if $\chi_f = \chi_g$ then f and g are identically distributed. (*Hint:* Let $h(x)$ be a Schwartz function and $\hat{h}(t)$ be its Fourier transform. Using the Fourier inversion formula, show that

$$\frac{1}{2\pi} \int_{-\infty}^{\infty} \chi_f(t)\hat{h}(t) \, dt = \int_X h(f) \, d\mu$$

and interpret the right-hand side as

$$\int_{\mathbf{R}} h \, dv_f$$

where v_f is the probability distribution of f.)

13. Let $(X, \mathcal{F}, \mu)$ be as in exercise 11. Show that, if f and g are *independent* random variables, then $\chi_{f+g} = \chi_f \chi_g$. (*Hint:* See equation 6 of §2.6.)

14. **a.** Let X be the unit interval, $\mathcal{F}$ the Borel sets, and μ Lebesgue measure. Show that, if $f = R_n$ is the nth Rademacher function, then $\chi_f = \cos t$.

   b. Let $S_n = \sum_{i=1}^n R_i$. Show that the characteristic function of S_n is $(\cos t)^n$.

   c. Using part b of exercise 11, show that

$$\int S_n^{2k} \, d\mu = (-1)^k \frac{d^{2k}}{dt^{2k}} [(\cos t)^n] \Big|_{t=0}$$

   (Compare with §1.1, exercise 18.)

15. Let $H = \sum_{n=1}^{\infty} (1/n) R_n$ be the randomized harmonic series. Let χ be its characteristic function. Show that

$$\chi(t) = \prod_{n=1}^{\infty} \cos\left(\frac{t}{n}\right)$$

§3.6 Some Applications of Fourier Series to Probability Theory

Let p_n, $-\infty < n < \infty$, be a sequence of nonnegative real numbers with the following three properties:

(1) $p_n = p_{-n}$

(2) $p_n = 0$ for all but finitely many n's

and

(3) $\sum p_n = 1$

We will consider in this section a generalized version of the random walk in which a point-mass moves randomly along the real line with transition probabilities p_{i-j}. To be more specific, suppose that, at time k, the position of the point-mass is the integer point i. At time $k + 1$ the point-mass is allowed to move to any integer position j for which p_{i-j} is nonzero and the probability of its moving to this position is assumed to be p_{i-j}. (For instance, if $p_1 = p_{-1} = \frac{1}{2}$

and the other p_k's are zero, the process we have just described is the usual random walk.) To normalize, we will assume that the position at time zero is the origin.

The basic random variables associated with this process are for each k,

(4)   the difference between the positions at time k and time $k + 1$

Notice that the probability distributions v associated with these random variables do not depend on k; in fact, they are all just the measure

$$(5) \qquad\qquad v(A) = \sum_{r \in A} p_r$$

for every Borel subset A of $\mathbf{R}$. Indeed, if A is the one-element set consisting of the integer r, then $v(A) = p_r$, the probability that the point-mass moves r units to the right (or left) at time k.

Now let $X = I$ be the unit interval, $\mathcal{F} = B_I$ the Borel sets, and $\mu = \mu_L$. In §2.6 we showed that there exist independent, identically distributed random variables $f_i : I \to \mathbf{R}$, $i = 1, 2, \ldots$, such that equation 5 is their common probability distribution. If we take $(X, \mathcal{F}, \mu)$ to model the sample space of the process described above and the f_k's to model the random variables described in display 4, it is clear that we get an adequate measure theoretic model of this process. In this model the sum

$$(6) \qquad\qquad S_n = \sum_{i=1}^{n} f_i$$

is *the position of the point-mass at time n.*

Let's consider the question of when and how often the point-mass returns to its initial position. In our model the probability that the point-mass returns to its initial position at time n is the measure of the set

$$\{\omega \in I; S_n(\omega) = 0\}$$

In the remainder of this section, we will show that

$$(7) \qquad\qquad \mu(\{S_n = 0; \text{i.o.}\}) = 1$$

That is, with probability one the point-mass returns infinitely often to its initial position. (Incidentally, we will assume from now on that the transition probability p_0 is less than one, for otherwise equation 7 is trivially true: The point-mass stays at the origin forever with probability one.)

The first step in the proof will be to get a simple description of the measure of the set where $S_n = 0$. This step will be done using Fourier series. Consider the sum

$$(8) \qquad\qquad g(t) = \sum_{n=-\infty}^{\infty} p_n e^{int}$$

By equation 2 this sum is finite; so there are no problems of convergence. We will show the following proposition.

Proposition 1. The measure of the set

$$\{\omega \in I;\ S_n(\omega) = r\}$$

is the rth Fourier coefficient of the function g^n.

Proof. We claim that for all k

(9) $$g(t) = E(e^{itf_k}) = \int_I e^{itf_k}\,d\mu$$

Indeed, display 4 tells us that f_k is an integer-valued function, taking on only a finite number of integer values. In addition, it tells us that the measure of the set

$$E_m = \{\omega \in I;\ f_k(\omega) = m\}$$

is just p_m. Thus the right-hand side of equation 9 is

$$\sum e^{itm}\mu(f_k = m) = \sum p_m e^{itm} = g(t)$$

as claimed. Because $f_1, \ldots, f_n$ are independent, so are $e^{itf_1}, \ldots, e^{itf_n}$; so

$$E(e^{itS_n}) = E(e^{itf_1} \times \cdots \times e^{itf_n}) = E(e^{itf_1}) \times \cdots \times E(e^{itf_n})$$

by equation 6 of §2.6. By equation 9 the right-hand side is $g(t)^n$. On the other hand, because S_n is also an integer-valued function, taking on only a finite number of integer values,

$$E(e^{itS_n}) = \int_I e^{itS_n}\,d\mu = \sum \mu(S_n = r)e^{itr}$$

Comparing the Fourier coefficients of this series with those of $g(t)^n$, we see that $\mu(S_n = r)$ is the rth Fourier coefficient of $g(t)^n$. □

Before continuing with the proof of equation 7, we point out a few properties of the function g that we will use in our proof:

(10) $\qquad\qquad\qquad\qquad g$ is real-valued,

(11) $\qquad\qquad g(0) = 1, \qquad g'(0) = 0, \qquad$ and $\qquad g''(0) < 0,$

(12) $\qquad\qquad\qquad\qquad |g(t)| \le 1,$

and

(13) $|g(t)| < 1 \qquad$ except at a finite number of points on the interval $[-\pi, \pi]$

Proof of properties 10–13: **By equation 1**

$$\overline{g(t)} = \sum p_n e^{-int} = \sum p_n e^{int} = g(t)$$

so g is real-valued. Differentiating equation 8,

$$g(0) = \sum p_n, \qquad g'(0) = i\sum n p_n, \qquad \text{and} \qquad g''(0) = -\sum n^2 p_n$$

By equation 3 the first sum is 1, and by equation 1 the second sum is zero. Finally, the third sum is negative, because, by equation 3, $p_n \neq 0$ for some $n \neq 0$ (because we are assuming that $p_0 \neq 1$). To prove inequalities 12 and 13, note that

$$g(t) = Reg(t) = \sum p_n \cos nt$$

so

$$|g(t)| \leq \sum p_n |\cos nt| \leq \sum p_n = 1$$

with equality holding if and only if $\cos nt = \pm 1$ whenever $p_n \neq 0$.   ▽

Using these facts we will prove the following proposition.

Proposition 2.   The sum

(14) $$\sum_{n=0}^{\infty} \mu(\{\omega \in I; S_n(\omega) = 0\})$$

is infinite.

Proof.   By proposition 1 this sum is identical to the sum

$$\frac{1}{2\pi} \sum \int_{-\pi}^{\pi} g^n \, d\mu = \lim_{N \to \infty} \left(\frac{1}{2\pi}\right) \int_{-\pi}^{\pi} \sum_{n=0}^{2N-1} g^n \, d\mu$$

The integrand in the integral on the right is nonnegative and monotone-increasing (why?) and converges to the limit

$$\frac{1}{1 - g(t)}$$

except at those points where $g(t) = -1$. Because the points with this property are finite in number, by equation 13, we get from the monotone convergence theorem

$$\sum_{n=0}^{\infty} \mu(S_n = 0) = \frac{1}{2\pi} \int_{-\pi}^{\pi} \frac{dt}{1 - g(t)}$$

Because the integrand on the right is nonnegative, we can show that the left-hand side is infinite by showing that

$$\int_{-\varepsilon}^{\varepsilon} \frac{dt}{1 - g(t)} = +\infty$$

for some $\varepsilon > 0$. Let C be a constant with

$$g''(0) > -2C$$

Because $g(0) = 1$ and $g'(0) = 0$, we get from Taylor's formula with remainder that

$$g(t) \geq 1 - Ct^2 > 0$$

on a small interval, $-\varepsilon < t < \varepsilon$. Hence

$$\int_{-\varepsilon}^{\varepsilon} \frac{dt}{1 - g(t)} \geq \int_{-\varepsilon}^{\varepsilon} \frac{dt}{Ct^2} = +\infty \qquad \square$$

If the events $S_n = 0$ were independent, we would now be finished: We could deduce that $\mu(\{S_n = 0; \text{ i.o.}\}) = 1$ from proposition 2 using the second Borel–Cantelli lemma. However, because these events are not independent, we have to resort to a slightly more complicated argument. Let k be a positive integer and, for every positive integer l, let

$$f_l' = f_{k+l}$$

Let B_k be the set of $\omega \in I$ where

$$(15) \qquad \sum_{i=1}^{r} f_i(\omega) \neq 0, \qquad r < k, \qquad \text{and} \qquad \sum_{i=1}^{k} f_i(\omega) = 0$$

Similarly, let B_l' be the set of $\omega \in I$ where

$$(16) \qquad \sum_{i=1}^{r} f_i'(\omega) \neq 0, \qquad r < l, \qquad \text{and} \qquad \sum_{i=1}^{l} f_i'(\omega) = 0$$

Because the f's in display 15 and the f''s in display 16 are independent, B_k and B_l' are independent.

Let $\rho_k = \mu(B_k)$ and $\rho_l' = \mu(B_l')$. We claim that, for $k = l$, $\rho_k = \rho_l'$. Indeed, let π be the joint probability distribution associated with $f_1, \ldots, f_k$, and let π' be the joint probability distribution associated with $f_1', \ldots, f_k'$. By theorem 3 of §2.6,

$$\pi = \mu_{f_1} \times \mu_{f_2} \times \cdots \times \mu_{f_k} = \overbrace{\nu \times \cdots \times \nu}^{k}$$

and a similar identity holds for π', so

$$\pi = \pi'$$

Now ρ_k is the measure with respect to μ of the set in display 15. But, by the definition of π, this measure is the same as the measure with respect to π of the set

$$\left\{(x_1,\ldots,x_n)\in\mathbf{R}^n;\ \sum_{i=1}^r x_i\neq 0,\ \ r<k,\ \ \text{and}\ \ \sum_{i=1}^k x_i=0\right\}$$

Similarly ρ_k' is the measure of this set with respect to π'. Because $\pi=\pi'$, $\rho_k=\rho_k'$ as claimed. By definition, B_k is the set of random paths that return for the first time to the origin at time k; so the sum

$$\rho=\sum_{k=1}^\infty \rho_k$$

is the probability that a random path returns *at least once* to the origin. Similarly $B_k\cap B_l'$ is the set of random paths that return to the origin for the first time at time k and for the second time at time $k+l$. Because B_k and B_l' are independent, the probability of this event—that is, the measure of $B_k\cap B_l'$—is $\rho_k\rho_l$; and the probability that a random path returns *at least twice* to the origin is

$$\sum_{k,l=1}^\infty \rho_k\rho_l=\left(\sum_{k=1}^\infty \rho_k\right)\left(\sum_{l=1}^\infty \rho_l\right)=\rho^2$$

We leave for the reader to show, by a similar argument, the following proposition.

Proposition 3. If ρ is the probability of a random path returning to its initial position at least once, the probability of its returning at least k times is ρ^k.

We will deduce from this proposition that ρ must equal 1. In fact, suppose that $\rho<1$. Let

$$A_k=\{\omega\in I;\ S_k(\omega)=0\}$$

and let

(17) $$h=\sum_{k=1}^\infty 1_{A_k}$$

Notice that, for $m<\infty$, the set

(18) $$\{\omega\in I;\ h(\omega)=m\}$$

is the set of random paths that return to the origin *exactly* m times; so, by proposition 3, the measure of this set is $\rho^m-\rho^{m+1}$ or $\rho^m(1-\rho)$. For $m=\infty$, the set in display 18 is the set of paths that return to the origin infinitely often; so, by the proposition, its measure is less than ρ^k for all k; in other words, it is zero (assuming that $\rho<1$). Thus

$$\int_I h\,d\mu = \sum_{m<\infty} m\rho^m(1-\rho) < \infty$$

On the other hand, by equation 17

$$\int_I h\,d\mu = \sum_{k=1}^{\infty} \mu(A_k)$$

and the quantity on the right is infinite, by proposition 2; so we get a contradiction and conclude that $\rho = 1$.

Let C_k be the set of paths that return to the origin at least k times. Because $\rho = 1$, $\mu(C_k) = 1$ by proposition 3. Because $C_1 \supset C_2 \supset C_3 \cdots$,

$$\mu\left(\bigcap_{i=1}^{\infty} C_k\right) = 1$$

Hence, we conclude that random paths return to the origin infinitely often with probability one.

Exercises for §3.6

1. Show that, for the classical random walk ($p_{-1} = p_1 = \frac{1}{2}$), every integer point $n \in Z$ is visited at least once with probability one. (*Hint:* Suppose that the random walk visits the integer point n with probability $p < 1$. If $n > 0$, consider the set of random walks having the following two properties: (i) The first n moves are to the left. (ii) The origin is never revisited. Prove that the probability that a random walk belongs to this set is $(1 - p)(1/2^n) > 0$, contradicting equation 7.)

2. Show that, for the classical random walk, every integer point is visited infinitely often. (*Hint:* Let q be the probability that the point n is visited at least once, and let p be the probability that the random walk returns at least once to the origin. Show that the probability that n is visited at least k times is qp^{k-1}. (Compare with proposition 3.) Now use the fact that $p = q = 1$.)

3. **a.** For the generalized random walk with transition probabilities satisfying equations 1 through 3, let $n_1, \ldots, n_k$ be those integers for which $p_n \neq 0$. Let A be the set of those integers that can be written in the form $r_1 n_1 + \cdots + r_k n_k$, with integers $r_1, \ldots, r_k$, and let n be the greatest common divisor of $n_1, \ldots, n_k$. Show that A consists of all integer multiples of n.

   b. We will call an integer point a on the real line *accessible* if $a \in A$. Show that a is visited with probability greater than zero if and only if it is accessible.

4. Show that, if an integer point a is accessible, then with probability one it is visited infinitely often by the generalized random walk. (Compare with exercises 1 and 2.)

5. a. Let $g(t)$ be the function given by equation 8. Show that $g(t) = 1$ if and only if t is a multiple of $2\pi/n$ (the n here being the same n as in exercise 3, part a).

   b. Conclude that $g(t) = 1$ if and only if $\cos kt = 1$ for all admissible k's.

   c. Show that if $g(t) = 1$ then

$$g''(t) = -\sum n^2 p_n$$

   Conclude that $g''(t) < 0$ when $g(t) = 1$.

6. Let

$$T_N(0) = \sum_{k=1}^{N} \text{Prob}(S_k = 0) \quad \text{and} \quad T_N(r) = \sum_{k=1}^{N} \text{Prob}(S_k = r)$$

Show that, if r is admissible, $\lim[T_N(0) - T_N(r)]$ is finite as $N \to \infty$ and is equal to

(∗)
$$\frac{1}{2\pi} \int_{-\pi}^{\pi} \frac{1 - \cos rt}{1 - g(t)} dt$$

(*Hint:* Use part b of exercise 5 to show that the integrand in equation (∗) is a bounded function of t.)

7. a. Show that, for the classical random walk ($p_{-1} = p_1 = \frac{1}{2}$),

$$\text{Prob}(S_{2n} = 0) = \frac{1}{4^n} \binom{2n}{n}$$

   b. By Stirling's formula (see S. Lang, *A Complete Course in Calculus.* Reading, Mass.: Addison–Wesley, 1968), there exists a number θ, with $0 < \theta < 1$, such that

$$n! = \sqrt{2\pi n}\, n^n e^{-n} e^{\theta/12n}$$

   Deduce from Stirling's formula that

$$\text{Prob}(S_{2n} \sim 0) = \frac{1}{\sqrt{\pi n}}$$

   for n large.

   c. Prove from part b that

$$\sum_{n=1}^{\infty} \text{Prob}(S_{2n} = 0) = \infty$$

   [Notice that $\text{Prob}(S_{2n+1} = 0) = 0$. Why?]

8. a. For the "unfair coin" [the process described in theorem 5 of §2.6 with $k = 2$, $r_1 = 1$, $r_2 = -1$, $p_1 = p \neq \frac{1}{2}$, and $p_2 = (1 - p) \neq \frac{1}{2}$], let $S_n = f_1 + \cdots + f_n$. Prove that

$$\text{Prob}(S_{2n} = 0) = \binom{2n}{n} p^n (1 - p)^n \quad \text{and} \quad \text{Prob}(S_{2n+1} = 0) = 0$$

b. Using Stirling's formula show that

$$\sum_{n=1}^{\infty} \text{Prob}(S_n = 0) < \infty$$

and conclude (by the first Borel–Cantelli lemma) that $\text{Prob}(S_n = 0;$ i.o.) $= 0$. Why doesn't this contradict equation 7?

9. Suppose that the transition probabilities p_n, $-\infty < n < \infty$, satisfy equations 1 and 3 but not equation 2. Show that if $\sum n^2 p_n < \infty$, proposition 2 is still valid.

10. a. For the generalized random walk with transition probabilities satisfying equations 1 through 3, let $N_n(k)$ be the number of times the point k is visited during the interval of time $0 \leq t \leq n$. Show that the expectation value of $N_n(k)$ is

$$\sum_{l=0}^{n} \text{Prob}(S_l = k)$$

(*Hint:* Let B_r be the set of paths that visit the point k exactly r times during the period $0 \leq t \leq n$. Show that

$$N_n(k) = \sum r 1_{B_r}$$

Let $A_l = \{\omega; S_l(\omega) = k\}$. Show that

$$N_n(k) = \sum_{l=0}^{n} 1_{A_l})$$

b. Show that the expectation value of $N_n(k)$ is

$$\frac{1}{2\pi} \int_{-\pi}^{\pi} \cos kt \, \frac{1 - g^{n+1}}{1 - g} \, dt$$

c. Using exercise 6 show that, if k is admissible, the expectation value of $N_n(k)$ differs from the expectation value of $N_n(0)$ by a quantity that tends to a finite limit as $n \to \infty$.

§3.7 An Application of Probability Theory to Fourier Series

In this section we are going to discuss a classical theorem about Fourier series due to G. Szegö. This theorem not only is of considerable theoretical interest, but also has a number of practical real-life applications. For those who want to learn about these applications, we recommend the very readable book by Grenander and Szegö (*Toeplitz Forms and Their Applications.* Berkeley, Calif.: University of California Press, 1958).

Szegö proved his theorem in 1916, and since then several other proofs of it have been discovered. The proof described below is due to Mark Kac and consists of reversing one of the key arguments of the previous section.

We begin by making a few definitions: Let $\mathscr{B}$ be the Borel subsets of the real line. Suppose we are given a probability measure μ and a sequence of probability measures $\mu_1, \mu_2, \ldots$ on $\mathscr{B}$. We will say that μ_n *converges weakly to* μ if, for every bounded continuous function f,

$$(1) \qquad \int f \, d\mu_n \to \int f \, d\mu$$

The notion of weak convergence will play an important role, not only in the following discussion, but also in the formulation of the central limit theorem in the next section.

The second notion we will need involves some elementary linear algebra. Let $T = (a_{ij})$, $1 \le i, j \le N$, be an $N \times N$ matrix of complex numbers with

$$(2) \qquad \bar{a}_{ij} = a_{ji}$$

It is a standard theorem in linear algebra that T has N real eigenvalues; that is, the equation

$$\det(\lambda - T) = 0$$

has N real roots $\lambda_1, \ldots, \lambda_N$ (potentially occurring with multiplicities). For every subset $A \in \mathscr{B}$, let $\mu(A)$ be the number of λ_i's contained in A, counting multiplicities. The measure μ defined by this recipe is called the *spectral measure* of T. If f is a bounded continuous function then

$$(3) \qquad \int f \, d\mu = \sum_{i=1}^{N} f(\lambda_i)$$

Having the notions of weak convergence and spectral measure, we can state a provisional form of the Szegö theorem. Let a_n, $-\infty < n < \infty$, be a sequence of complex numbers satisfying

$$(4) \qquad a_{-n} = \overline{a_n}$$

and

$$(5) \qquad \sum_{n=-\infty}^{\infty} |a_n| < \infty$$

Associated with these numbers is the infinite matrix

$$T = (a_{mn}) \quad \text{with} \quad a_{mn} = a_{m-n} \qquad -\infty < m, \quad n < \infty$$

Matrices of this form are called *Toeplitz* matrices. Notice that, by equation 4,

T satisfies the symmetry condition in equation 2. Let

$$(6) \qquad\qquad T_N = (a_{m-n}) \qquad 0 \le m, n \le N - 1$$

be the $N \times N$ principal minor of this matrix. The Szegö theorem in its provisional form says that, if μ_N is the spectral measure of T_N, then μ_N / N has a weak limit as $N \to \infty$. We will identify this limit shortly, but before doing so let's observe an interesting tie-in between Toeplitz matrices and Fourier series. Let

$$(7) \qquad\qquad q(\theta) = \sum_{n=-\infty}^{\infty} a_n e^{in\theta}$$

By equation 5, q is continuous. Moreover, by equation 4

$$\bar{q}(\theta) = \sum \bar{a}_{-n} e^{in\theta} = \sum a_n e^{in\theta} = q(\theta)$$

That is, q is real-valued. Let $\mathscr{L}^2$ be the space of $\mathscr{L}^2$-integrable functions on the interval $[-\pi, \pi]$ and let

$$T_q : \mathscr{L}^2 \to \mathscr{L}^2$$

be the linear mapping that sends $f \in \mathscr{L}^2$ to $qf \in \mathscr{L}^2$. Then, by equation 7

$$(8) \qquad\qquad T_q e^{im\theta} = \sum_{n=-\infty}^{\infty} a_n e^{i(n+m)\theta} = \sum_{n=-\infty}^{\infty} a_{n-m} e^{in\theta}$$

That is, a_{n-m} is the matrix associated with T_q in terms of the basis $e^{in\theta}$, $-\infty < n < \infty$, of $\mathscr{L}^2$. The matrix T_N has a similar description: Let V_N be the vector subspace of $\mathscr{L}^2$ spanned by the functions $e^{in\theta}$, $0 \le n \le N - 1$, and let

$$P : \mathscr{L}^2 \to V_N$$

be the orthogonal projection of $\mathscr{L}^2$ onto V_N. In other words, if f is in $\mathscr{L}^2$ and its Fourier series is

$$\sum_{-\infty}^{\infty} c_n e^{in\theta}$$

then

$$Pf = \sum_{0}^{N-1} c_n e^{in\theta}$$

Let

$$(T_q)_N : V_N \to V_N$$

be the linear mapping

$$(9) \qquad\qquad (T_q)_N f = PT_q f$$

Then, for $0 \leq m \leq N - 1$,

$$(T_q)_N e^{im\theta} = \sum_{n=0}^{N-1} a_{n-m} e^{in\theta}$$

by equation 8; so the matrix associated with $(T_q)_N$ in terms of the basis $e^{in\theta}$, $0 \leq n \leq N - 1$, is exactly T_N.

Consider the probability measure

$$\left(\frac{1}{2\pi} \right) \mu_L$$

on the interval $[-\pi, \pi]$. If we think of the function $q : [-\pi, \pi] \to \mathbf{R}$ as a random variable, its probability distribution μ is defined by the formula in equation 3 of §2.6; that is, for every Borel function f,

(10)
$$\int_{\mathbf{R}} f d\mu = \frac{1}{2\pi} \int_{-\pi}^{\pi} f[q(\theta)] d\mu_L$$

We can now state the Szegö theorem in its sharp form.

Theorem 1.   Let μ_N be the spectral measure of T_N, and let μ be the measure in equation 10. Then

(11)
$$\frac{\mu_N}{N} \to \mu$$

weakly as $N \to \infty$.

Let us see what equation 11 says in concrete terms: Let

$$\lambda_i^{(N)} \qquad i = 1, \ldots, N$$

be the eigenvalues of T_N, and let f be a bounded continuous function on the real line. Then by equation 3

$$\int f d\mu_N = \sum_{i=1}^{N} f(\lambda_i^{(N)})$$

so equation 11 is equivalent to the assertion that

(12)
$$\frac{1}{N} \sum_{i=1}^{N} f(\lambda_i^{(N)}) \to \frac{1}{2\pi} \int_{-\pi}^{\pi} f[q(\theta)] d\theta$$

as $N \to \infty$, for every bounded continuous function f.

We will now give a heuristic justification of equation 12 in terms of probability theory. Suppose that

$$a_n = p_n \qquad -\infty < n < \infty$$

with the p_n's being transition probabilities satisfying equations 1 through 3 of §3.6. As in §3.6 we will consider the random walk based on these transition probabilities; but, instead of assuming that the random path starts at the origin, we will assume that it starts at one of the points $x = 0, \ldots, N - 1$ and that all these points are equally likely as starting points. We will confine ourselves to a fixed finite interval of time $0 \le t \le m$. Thus a sample path can be described as a zigzag line consisting of m segments. An example of a path for $m = 3$ is shown in the following figure. The t coordinate indicates time, and the x coordinate indicates position.

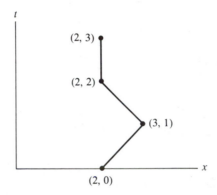

We will compute the return-time probabilities for this process as in §3.6 but with a "confinement" condition imposed: What is the probability that a random path returns to its initial position at time $t = m$ *and* stays confined in the box $0 \le x \le N - 1$ and $0 \le t \le m$? (For instance, in the figure the confinement condition says that, at times $t = 0$, $t = 1$, $t = 2$, and $t = 3$, the x coordinate of the path has to lie in the interval $0 \le x \le N - 1$.) Let us denote the probabilities in question by $p(m, N)$. We claim that

$$(13) \qquad\qquad p(m, N) = \frac{1}{N} \sum_{i=1}^{N} (\lambda_i^{(N)})^m$$

Proof. We will prove the case of $m = 3$, the general case being essentially no more difficult. Let k be a point on the interval $0 \le x \le N - 1$. Because our random path has to start at some point in this interval and because all N points are equally likely, the probability that it starts at k is $1/N$. What is the probability that at $t = 0$ its position is k, at $t = 1$ its position is l, at $t = 2$ its position is m, and at $t = 3$ its position is again k? Clearly this is

$$\left(\frac{1}{N}\right) p_{l-k} p_{m-l} p_{k-m}$$

The probability $p(3, N)$ is therefore the sum

(14)
$$\left(\frac{1}{N}\right) \sum p_{l-k} p_{m-l} p_{k-m}$$

over $0 \le k, l, m \le N - 1$. But the sum in display 14 is also just $1/N$ times the trace of the matrix T_N^3; that is,

(15)
$$p(3, N) = \left(\frac{1}{N}\right) \text{trace } T_N^3$$

In terms of the eigenvalues $\lambda_i^{(N)}$, this trace is just

$$\sum_{i=1}^{N} (\lambda_i^{(N)})^3$$

establishing equation 13 for $m = 3$. $\qquad\qquad\qquad\qquad\qquad\qquad\qquad\qquad$ $\square$

Let's now drop the constraint condition—that is, no longer require the random path to be constrained to lie on the interval $0 \le x \le N - 1$ but only require that its initial position lie on this interval. What is the probability, for an unconstrained random path, that its positions at time $t = 0$ and at time $t = m$ coincide? Clearly this is just the "return at time m" probability computed in §3.6; that is, it is just

$$p(m) = \frac{1}{2\pi} \int_{-\pi}^{\pi} q(\theta)^m \, d\theta$$

where $q(\theta) = \sum p_n e^{in\theta}$ by proposition 1 of §3.6. Now it is intuitively clear that as $N \to \infty$, with m fixed, $p(m, N) \to p(m)$. Indeed, if we make the interval $[0, N - 1]$ extremely large relative to m, relatively few paths with an initial point in this interval leave it before $t = m$ in view of property 2 of §3.6. Hence, we conclude

(16)
$$\frac{1}{N} \sum_{i=1}^{N} f(\lambda_i^{(N)}) \to \frac{1}{2\pi} \int_{-\pi}^{\pi} f[q(\theta)] \, d\theta$$

for $f(x) = x^m$. By taking linear combinations of x^m's we see that equation 16 is true for any polynomial function f, and a simple application of the Weierstrass approximation theorem shows that it is true in general.

The following series of exercises will give you a chance to strip the probabilistic scaffolding from this proof. Henceforth, a_n, $-\infty < n < \infty$, will be an arbitrary sequence of complex numbers satisfying equations 4 and 5; that is, the a_n's will not necessarily be the transition probabilities associated with a random walk.

1. Show that

$$\frac{1}{2\pi} \int_{-\pi}^{\pi} q^3 \, d\theta = \sum_{r+s+t=0} a_r a_s a_t$$

2. Suppose that all but finitely many of the a_n's are zero. Show that

$$\left(\frac{1}{N}\right) \text{trace } T_N^3 \to \frac{1}{2\pi} \int_{-\pi}^{\pi} q^3 \, d\theta$$

3. Let a'_n, $-\infty < n < \infty$, be another sequence of numbers satisfying equations 4 and 5. Suppose that, for all n, $|a_n - a'_n| < \varepsilon$. Let M be the larger of the two sums

$$\sum |a_n| \quad \text{and} \quad \sum |a'_n|$$

Show that

$$\left| \sum_{r+s+t=0} a_r a_s a_t - \sum_{r+s+t=0} a'_r a'_s a'_t \right| < 3\varepsilon M^2$$

4. Similarly, show that

(*) $$\left(\frac{1}{N}\right) |\text{trace } T_N^3 - \text{trace} (T'_N)^3| < 3\varepsilon M^2$$

5. Show that equation (*) is true without the assumption that all but finitely many of the a_n's are zero.
6. Prove equation (*) with the power 3 replaced by the power m.
7. Let $a = \min q$ and $b = \max q$. Show that the eigenvalues of T_N lie on the interval $a \le \lambda \le b$. (*Hint:* Use equation 9.)
8. Let f be a continuous function on the interval $[a, b]$. By the Weierstrass approximation theorem, there exists for every $\varepsilon > 0$ a polynomial function p such that $\sup |f - p| < \varepsilon$. (See §2.7, exercise 9, or §3.4 exercise 6.) Use this theorem in conjunction with exercise 7 to prove that μ_N converges weakly to μ.
9. (Szegö's original version of the Szegö theorem) Let D_N be the determinant of T_N. Show that, if q is bounded below by a positive number, then $D_N > 0$ and

$$\lim_{N \to \infty} D_N^{1/N} = \exp\left[\frac{1}{2\pi} \int_{-\pi}^{\pi} \log q(\theta) \, d\theta\right]$$

§3.8 The Central Limit Theorem

In §2.7 we proved the following version of the law of large numbers.

Theorem 1. Let $f_1, f_2, \ldots$ be a sequence of bounded random variables on X that are independent and identically distributed. Let $E = E(f_i)$ be the common expectation value of the f_i's. Let X_0 be the set of points $x \in X$ for which

(1)
$$\frac{f_1(x) + \cdots + f_n(x)}{n} \to E$$

as $n \to \infty$. Then $\mu(X_0) = 1$.

In other words, if

$$S_n(x) = f_1(x) + \cdots + f_n(x) - nE$$

then

(2)
$$\frac{S_n(x)}{n} \to 0$$

as $n \to \infty$ with probability one. In many practical problems, one would like to know how fast this convergence is. A close look at the proof of theorem 1 gives the following theorem.

Theorem 2.   Let $f_1, f_2, \ldots$ be as in theorem 1, and let $\alpha > 0$ be given. Then

(3)
$$\frac{S_n(x)}{n^{(1/2)+\alpha}} \to 0$$

as $n \to \infty$ with probability one; that is, $S_n(x)/n \to 0$ faster than $n^{-(1/2)+\alpha}$ with probability one.

Proof.   The first step of the proof is to verify the following lemma.

Lemma 3.   For each $k > 0$ there is a $C_k > 0$ such that

(4)
$$\int_X S_n^{2k} \, d\mu \leq C_k n^k$$

This lemma can be proved by induction. The case of $k = 2$ is done in the proof of theorem 1. The induction step is fairly messy but straightforward, and we leave it to the reader. (See §2.7, exercise 5.)

Now, given $\alpha > 0$, choose k so that $1/k < \alpha$; notice that equation 2 and Chebyshev's inequality give

$$\mu\left(\left\{\left|\frac{S_n}{n^{(1/2)+\alpha}}\right| > \varepsilon\right\}\right) = \mu\left(\left\{\frac{S_n^{2k}}{n^{k+2k\alpha}} > \varepsilon^{2k}\right\}\right)$$

(5)
$$\leq \frac{1}{\varepsilon^{2k} n^{k+2k\alpha}} \int_X S_n^{2k} \, d\mu$$

$$\leq \frac{C_k}{\varepsilon^{2k} n^{2k\alpha}} \leq \frac{C_k}{\varepsilon^{2k} n^{2+\beta}}$$

where $\beta = 2k\alpha - 2 = 2k(\alpha - 1/k) > 0$.

Now, as in §1.1, choose a sequence $\varepsilon_1, \varepsilon_2, \ldots$ with $\varepsilon_n \to 0$ and

$$\sum_{n=1}^{\infty} \frac{C_k}{\varepsilon_n^{2k} n^{2+\beta}} < \infty$$

Let

$$A_n = \left\{ x \in X; \left| \frac{S_n(x)}{n^{(1/2)+\alpha}} \right| > \varepsilon_n \right\}$$

Then, by equation 5, $\sum_{n=1}^{\infty} \mu(A_n) < \infty$. So, by the first Borel–Cantelli lemma, $\mu(\{A_n; \text{i.o.}\}) = 0$. But, for x in the complement of $\{A_n; \text{i.o.}\}$, we must have

$$\left| \frac{S_n(x)}{n^{(1/2)+\alpha}} \right| < \varepsilon_n$$

for all but finitely many n's. Hence, we conclude

$$\frac{S_n(x)}{n^{(1/2)+\alpha}} \to 0 \quad \text{as} \quad n \to \infty$$

with probability one.   □

A natural question to ask at this point is "What happens to $S_n/n^{1/2}$ as $n \to \infty$?" To formulate the answer to this question, we need a definition. Let f be a bounded random variable with expectation value E. The integral

$$V(f) = \int_X (f - E)^2 \, d\mu$$

is called the *variance* of f. It is regarded by probabilists as a good measure of "deviation of f from its expectation value," because by Chebyshev's inequality,

(6) $$\mu\{x \in X; |f - E| > M\} \le \frac{1}{M^2} \int (f - E)^2 \, d\mu = \frac{V(f)}{M^2}$$

For instance, inequality 6 says that the set where f deviates by one unit from its expectation value is less than $V(f)$. If $V(f)$ is very small, so is this deviation.

Notice that, if v is the probability distribution associated with f, then by equation 3 of §2.6

$$V(f) = \int_R (x - E)^2 \, dv$$

So, if two random variables are identically distributed, they have the same variance.

Let's now return to the question posed earlier. Because $f_1, f_2, \ldots$ are identically distributed, $V(f_1) = V(f_2) = \cdots$.

Theorem 4. Let $\sigma = V(f_1) = V(f_2) = \cdots$. Then, for every pair of numbers a and b, with $a < b$

(7)
$$\mu\left\{x \in X; a < \frac{S_n(x)}{n^{1/2}} < b\right\} \to \frac{1}{\sqrt{2\pi\sigma}} \int_a^b e^{-t^2/2\sigma} \, dt$$

as n tends to infinity.

This theorem is called the *central limit theorem*. It is sometimes stated as saying that, if the deviations of the f_i's from their expectation value E, for $1 \le i \le n$, are rescaled by the factor $n^{1/2}$, then these deviations tend to be *normally distributed* for n large.

Notice that, if we denote by μ_n the probability distribution of the random variable $S_n/n^{1/2}$ and by J the interval (a, b), the left-hand side of equation 7 is $\mu_n(J)$. The right-hand side, on the other hand, is

(8)
$$\mu_\sigma(J) = \frac{1}{\sqrt{2\pi\sigma}} \int_J e^{-t^2/2\sigma} \, d\mu_L$$

The measure μ_σ defined by equation 8 is called the *Gaussian* or *normal distribution with variance* σ. The central limit theorem says that, for every interval J,

(9)
$$\mu_n(J) \to \mu_\sigma(J) \quad \text{as} \quad n \to \infty$$

Proof of theorem 4. Replacing f_i by $f_i - E$, we can assume that $E = 0$. We will first prove a statement similar to equation 9 for the Fourier transforms of μ_n and μ_σ.

Lemma 5. Let

$$\chi_n(t) = \int_{\mathbf{R}} e^{-ixt} \, d\mu_n$$

Then, for fixed t, $\chi_n(t) \to e^{-\sigma t^2/2}$ as $n \to \infty$.

Proof. By equation 3 of §2.6,

$$\chi_n(t) = \int_{\mathbf{R}} e^{-ixt} \, d\mu_n = \int_X e^{-it(S_n/\sqrt{n})} \, d\mu$$

$$= \int_X e^{-(it/\sqrt{n})(f_1 + \cdots + f_n)} \, d\mu$$

$$= \left(\int_X e^{-itf_1/\sqrt{n}} \, d\mu\right) \times \cdots \times \left(\int_X e^{-itf_n/\sqrt{n}} \, d\mu\right)$$

by independence. Because the f_i's are identically distributed, equation 3 of §2.6 tells us that

$$\int_X e^{-itf_1/\sqrt{n}} \, d\mu = \int_X e^{-itf_2/\sqrt{n}} \, d\mu = \cdots = \int_X e^{-itf_n/\sqrt{n}} \, d\mu$$

In fact, each of these expressions is equal to

$$\int_{\mathbf{R}} e^{-itx/\sqrt{n}} \, dv$$

where v is the common probability distribution of the f_i's. Therefore, letting f be any one of the f_i's, we get for χ_n the formula

(10)
$$\chi_n(t) = \left(\int_X e^{-itf/\sqrt{n}} \, d\mu \right)^n$$

Now notice that, for t fixed,

$$e^{-itf/\sqrt{n}} = 1 - \frac{itf}{\sqrt{n}} + \left(\frac{1}{2!} \right) \left(\frac{itf}{\sqrt{n}} \right)^2 - \left(\frac{1}{3!} \right) \left(\frac{itf}{\sqrt{n}} \right)^3 + \cdots$$

$$= 1 - \frac{itf}{\sqrt{n}} - \left(\frac{t^2}{2n} \right) f^2(1 + r_n)$$

where r_n is a bounded function that tends to zero uniformly as $n \to \infty$. Integrating the right-hand side, taking into account the fact that $E = \int f \, d\mu = 0$, we get

(11)
$$\int_X e^{-itf/\sqrt{n}} \, d\mu = 1 - \left(\frac{\sigma t^2}{2n} \right)(1 + \varepsilon_n)$$

where

$$\varepsilon_n = \frac{1}{\sigma} \int_X f^2 r_n \, d\mu$$

The Lebesgue dominated convergence theorem implies that

(12)
$$\varepsilon_n \to 0 \quad \text{as} \quad n \to \infty$$

If we substitute equation 11 into equation 10, we get

$$\chi_n(t) = \left[1 - \left(\frac{\sigma t^2}{2n} \right)(1 + \varepsilon_n) \right]^n$$

Therefore, the proof of lemma 5 reduces to showing that

(13)
$$\left[1 - \left(\frac{\sigma t^2}{2n}\right)(1 + \varepsilon_n)\right]^n \to e^{-\sigma t^2/2}$$

as $n \to \infty$. For this proof, set $a = -\sigma t^2/2$ and take the log of both sides. The left-hand side becomes

$$n \log\left[1 + \left(\frac{a}{n}\right)(1 + \varepsilon_n)\right]$$

or, with $s = (1 + \varepsilon_n)/n$,

$$\frac{1}{1 + \varepsilon_n} \frac{\log(1 + as)}{s}$$

In view of equation 12, the limit of this expression as n tends to infinity is

$$\lim_{s \to 0} \frac{\log(1 + as)}{s} = \frac{d}{ds} \log(1 + as)|_{s=0} = a$$

which is exactly the log of the right-hand side of equation 13.   ▽

We now return to the proof of theorem 4. We begin by proving an essentially equivalent statement. We will show that, if f is a Schwartz function, then

(14)
$$\int_{\mathbf{R}} f d\mu_n \to \int_{\mathbf{R}} f d\mu_\sigma$$

as $n \to \infty$. To see this, note that

$$\int_{\mathbf{R}} f d\mu_n = \frac{1}{2\pi} \int_{\mathbf{R}} \left(\int_{\mathbf{R}} \check{f}(t) e^{-ixt} dt\right) d\mu_n$$

$$= \frac{1}{2\pi} \int_{\mathbf{R}} \check{f}(t) \chi_n(t) dt$$

The last expression, however, limits to

$$\frac{1}{\sqrt{2\pi\sigma}} \int_{\mathbf{R}} f(y) e^{-y^2/2\sigma} dy$$

by lemma 4 of §3.5 and lemma 5 of this section. Hence, we have established equation 14.

To show that equation 14 is essentially equivalent to equation 9, consider a Schwartz function f_ε of the form indicated in the following figure. (We showed in §3.5 that Schwartz functions of this type do exist.)

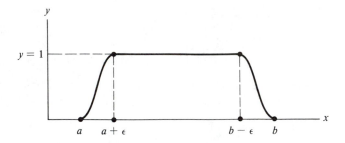

Let J be the interval (a, b). It is clear that

$$\int f_\varepsilon \, d\mu_n \leq \mu_n(J)$$

so by equation 14

$$\liminf \mu_n(J) \geq \lim_{n \to \infty} \int f_\varepsilon \, d\mu_n = \int f_\varepsilon \, d\mu_\sigma$$

$$\geq \int_{a+\varepsilon}^{b-\varepsilon} e^{-x^2/2\sigma} \, dx$$

The last inequality holds for all $\varepsilon > 0$, so we obtain

$$\liminf \mu_n(J) \geq \int_a^b e^{-x^2/2\sigma} \, dx = \mu_\sigma(J)$$

A similar argument shows that

$$\limsup \mu_n(J) \leq \mu_\sigma(J)$$

so we conclude that the limit exists and is equal to the expression on the right.   □

   Another formulation of the central limit theorem is that the *sequence of measures μ_n converges weakly to the measure μ_σ as $n \to \infty$*. Recall from §3.7 that this statement means that

(15)
$$\int_{\mathbf{R}} f \, d\mu_n \to \int_{\mathbf{R}} f \, d\mu_\sigma$$

for every bounded continuous function f. By equation 14 we know this fact to be true when f is a Schwartz function; and, by approximating by Schwartz functions, one can easily show it to be true for any bounded continuous function.

Example: coin tossing.   Suppose that in theorem 4 the f_i's are the Rademacher functions R_i. The strong law of large numbers says that $(S_n/n)(\omega) \to 0$ as

$n \to \infty$. A gambler might want to know how many trials it takes to be reasonably sure this quantity is near zero. For example he or she might want to know that $|S_n|/n < .01$ with a probability of 99%. Because $\sigma = V(R_1) = V(R_2) = \cdots = 1$, one gets from equation 7 the estimate

$$.99 = \mu\left(\left\{x \in X; \frac{|S_n|}{n} < .01\right\}\right)$$

(16)
$$= \mu\left(\left\{x \in X; \frac{|S_n|}{\sqrt{n}} < .01\sqrt{n}\right\}\right)$$

$$\approx \frac{1}{\sqrt{2\pi}} \int_{-.01\sqrt{n}}^{.01\sqrt{n}} e^{-t^2/2} \, dt$$

By numerical methods one can show that, if

$$\frac{1}{\sqrt{2\pi}} \int_{-a}^{a} e^{-t^2/2} \, dt = .99$$

then $a = 2.57\ldots$. Hence, by equation 16, $.01\sqrt{n} \approx 2.57$ or $n \approx 66,000$; that is, after 66,000 tosses one can be 99% sure that $|S_n|/n < .01$.

Appendix A
Metric Spaces

We collect here, in a minimal sense, the important facts about metric spaces used in the text. This exposition is in no way complete and is meant only as an easy reference for the reader who is already familiar with these concepts. For a more thorough treatment of this matter, see W. Rudin, *Principles of Mathematical Analysis, 3rd Ed.* (New York: McGraw-Hill, 1976).

Let M be a set.

Definition 1. A metric on a set M is a map $d(\cdot, \cdot): M \times M \to \mathbf{R}$ satisfying the properties

1. $d(x, y) = d(y, x)$
2. $d(x, y) \geq 0$
3. $d(x, y) = 0$ if and only if $x = y$
4. $d(x, y) \leq d(x, z) + d(z, y)$

If d is a metric on M, the pair (M, d) is called a *metric space*.

Example 2. Let $(V, \|\cdot\|)$ be a normed vector space (see §3.1). Then V is a metric space with metric

$$d(x, y) = \|x - y\|$$

Metric spaces are nice because they allow us to define the basic topological objects we are used to considering in $\mathbf{R}^n$—for example, open and closed sets, compactness, convergence, and so on. We first discuss convergence.

Definition 3. Let (M, d) be a metric space. Let $x_1, x_2, x_3, \ldots$ be a sequence in M. We say $\{x_n\}$ is a *Cauchy sequence* if

(1) $$d(x_n, x_m) \to 0 \quad \text{as} \quad n, m \to \infty$$

We say x_n *converges* to $x \in M$ (written $x_n \to x$), if

(2) $$d(x, x_n) \to 0 \quad \text{as} \quad n \to \infty$$

Proposition 4. If $x_1, x_2, \ldots$ is a sequence in M with $x_n \to x \in M$, then the x_i's form a Cauchy sequence.

Proof. $d(x_n, x_m) \leq d(x_n, x) + d(x, x_m) \to 0 \quad \text{as} \quad n, m \to \infty$ $\qquad \square$

Thus, just as in $\mathbf{R}^n$, every convergent sequence in (M, d) is a Cauchy sequence. The converse, however, is not true in general.

Example 5. Let $M = \mathbf{R} - \{0\}$ and let $d(x, y) = |x - y|$ for $x, y \in M$. Clearly the sequence $x_n = 1/n$ is Cauchy, and yet there is no $x \in M$ such that $x_n \to x$.

Experience in $\mathbf{R}^n$ tells us that it is nice to be able to use display 1 as a criterion for convergence. This motivates the following definition.

Definition 6. (M, d) is called *complete* if every Cauchy sequence in M converges to an element of M.

Example 7. $\mathbf{R}^n$ with the metric

(3) $$d(x, y) = \left(\sum_{j=1}^{n} (x_i - y_i)^2 \right)^{1/2}$$

is a complete metric space. We leave this fact for the reader to check. It follows from the fact that $\mathbf{R}$ with the metric in example 5 is complete.

When we consider convergence of sequences, it is sometimes useful to know if a subsequence converges. One check for this is that the sequence be contained in a compact set. To define compactness in a metric space, we need to study the open and closed sets.

Let (M, d) be a metric space and let $A \subset M$ be a subset. We say $x \in M$ is a *limit point* of A if for every $\varepsilon > 0$ there is an $x_\varepsilon \in A$ with $x_\varepsilon \neq x$ and $d(x, x_\varepsilon) < \varepsilon$. The set A is called *closed* if it contains all its limit points. In general, if $A \subset M$, the *closure* of A, $\bar{A}$, is the smallest closed set containing A. It is easy to check that $\bar{A}$ consists of the points in A together with all the limit points of A. A set $U \subset M$ is called *open* if its complement is closed.

Proposition 8. If $U \subset M$ is open, then for each $x_0 \in U$ there is an $r > 0$ such that the ball

(4) $$B_r(x_0) = \{x \in M; \, d(x, x_0) < r\}$$

is contained in U.

Proof. By definition, U is open if and only if U^c is closed. Now $x_0 \notin U^c$, so it is not a limit point of U^c. Hence there exists an $r > 0$ such that $B_r(x_0) \cap U^c = 0$. $\qquad\qquad\qquad\qquad\qquad\qquad\qquad\qquad\qquad\qquad\qquad\square$

A set $K \subset M$ is called *compact* if, whenever $\{U_\alpha\}$ is a collection of open sets covering K, there is a finite subcollection $U_{\alpha_1}, U_{\alpha_2}, \ldots, U_{\alpha_n}$ covering K. (A collection $\{U_\alpha\}$ of sets *covers* K if $K \subset \bigcup U_\alpha$.)

Proposition 9. Let (M, d) be a metric space and let $x_1, x_2, \ldots$ be an infinite sequence in M. Suppose there is a compact set $K \subset M$ such that $x_n \in K$ for all n. Then there is a subsequence $x_{n_1}, x_{n_2}, \ldots$ of the x_n's that converges.

Proof. It is enough to show that the set $A = \{x_1, x_2, \ldots\}$ has a limit point. Suppose this is not true. Then for each $y \in K$ there is an ε_y such that, if $B_y = \{x; \, d(x, y) < \varepsilon_y\}$, then B_y contains at most one point of A. The collection $\{B_y\}_{y \in K}$ is an open cover of K with no finite subcover. This fact contradicts the compactness of K. $\qquad\qquad\qquad\qquad\qquad\qquad\qquad\qquad\qquad\square$

This proposition gives a nice criterion for existence of convergent subsequences. The rub is that compactness is generally hard to check. However, in the case that the metric space is $\mathbf{R}^n$ with the usual metric given by equation 3, the Heine–Borel theorem gives a simple criterion for compactness.

Theorem 10. (Heine–Borel) Every closed, bounded subset of $\mathbf{R}^n$ is compact.

Remark. A set $A \subset \mathbf{R}^n$ is *bounded* if there is a number $M > 0$ such that $d(x, y) < M$ for all $x, y \in A$.

Warning. The Heine–Borel theorem is not true in a general metric space. To prove theorem 10 we need the following lemma.

Lemma 11. Let $(a_1, a_2, \ldots, a_n)$ and $(b_1, \ldots, b_n)$ be elements in $\mathbf{R}^n$ with $a_j \leq b_j$ for all $j = 1, 2, \ldots, n$. Then the closed multi-interval $J = \{x \in \mathbf{R}^n; \, a_i \leq x_i \leq b_i\}$ is compact.

Proof. Suppose, on the contrary, that J is not compact. Then there must be a cover of J by open sets $\{U_\alpha\}$ that has no finite subcover.

If we let $I_i \subset \mathbf{R}$ be the interval $I_i = [a_i, b_i]$, we note that $J = I_1 \times I_2 \times \cdots \times I_n$. Let $c_i = \frac{1}{2}(a_i + b_i)$ and let

$$I_i^- = [a_i, c_i] \qquad I_i^+ = [c_i, b_i] \qquad i = 1, \ldots, n$$

Then there are 2^n multi-intervals of the form

$$I_1^{\pm} \times I_2^{\pm} \times \cdots \times I_n^{\pm}$$

all of which are covered by the U_α's. At least one of these multi-intervals must not allow a finite subcover because J doesn't. Choose one such multi-interval and call it J_1. Now repeat this process ad infinitum to get a sequence of multi-intervals

$$J \supset J_1 \supset J_2 \supset \cdots$$

none of which can be covered by finitely many of the U_α's.

For each k take $x_k \in J_k$. Notice that the sequence $\{x_k\}_{k=1}^{\infty}$ is a Cauchy sequence because the size of J_k decreases as $k \to \infty$. Because J_1 is closed there is a limit point $x_0 \in J_1$. Also, for each l the tail of the sequence $\{x_k\}_{k=l}^{\infty}$ is in the closed set J_l. Hence $x_0 \in J_l$ for each l; that is, $x_0 \in \bigcap_{l=1}^{\infty} J_l$. Choose α_0 so that $x_0 \in U_{\alpha_0}$. By proposition 8 there is an $r > 0$ such that $B_r(x_0) \subset U_{\alpha_0}$. We claim that, for some k, $J_k \subset B_r(x_0)$. This claim contradicts the construction of the J_k's and thus proves the lemma. To prove this claim let

$$\lambda = \left(\sum_{j=1}^{n} (a_j - b_j)^2 \right)^{1/2}$$

and note that for x, $y \in J_k$, $d(x, y) \le \lambda/2^k$. Now, $x_0 \in J_k$ for all k, so we have that, if $x \in J_k$, then $d(x, x_0) \le \lambda/2^k$; that is, $J_k \subseteq B_{\lambda/2^k}(x_0)$. Choosing k large enough so that $\lambda/2^k < r$ we are finished. $\qquad\triangledown$

Proof of theorem 10. Let C be a closed, bounded subset of $\mathbf{R}^n$. Because C is bounded, it is contained in some closed multi-interval J. Because C is closed, the set $\mathbf{R}^n - C = U_0$ is open. Now let $\{U_\alpha\}$ be an open cover of C; the collection $\{U_\alpha\} \cup \{U_0\}$ is then an open cover of J. By the lemma, J is compact so there is a finite subcover. If U_0 is among these, throw it out; what's left still covers C. $\qquad\square$

Finally, we can combine proposition 9 with theorem 10 to get the Bolzano–Weierstrass property.

Theorem 12. (Bolzano–Weierstrass) Every bounded infinite set in $\mathbf{R}^n$ has a limit point.

Now suppose that (M, d_M) and (N, d_N) are two metric spaces. A function $f: M \to N$ is *continuous at* $x_0 \in M$ if for every $\varepsilon > 0$ there exists $\delta > 0$ such that $d_M(x_0, y) < \delta$ implies that $d_N(f(x_0), f(y)) < \varepsilon$. The function f is called *continuous* if it is continuous at each point of M. Notice that when checking continuity the δ may vary depending on x_0. The function f is called *uniformly continuous* if the δ can be chosen independently of x_0; namely, f is uniformly continuous if for every $\varepsilon > 0$ there is a $\delta > 0$ such that $d_N(f(x), f(y)) < \varepsilon$ whenever $d_M(x, y) < \delta$.

Proposition 13.   Let $f : M \to N$ be uniformly continuous and let $x_1, x_2, \ldots$ be a Cauchy sequence in M. Then $f(x_1), f(x_2), \ldots$ is a Cauchy sequence in N.

Proof.   Given $\varepsilon > 0$ we need to find K such that $d_N(f(x_i), f(x_j)) < \varepsilon$ whenever $i, j > K$. Because f is uniformly continuous, there is a $\delta > 0$ such that $d_M(x, y) < \delta$ implies that $d_N(f(x), f(y)) < \varepsilon$. Because the x_i's are a Cauchy sequence, we can find K such that $d_M(x_i, x_j) > \delta$ whenever $i, j < K$. This is the K we sought. $\square$

Now let (M, d_M) be a metric space and let $A \subset M$ be a subset. A is automatically a metric space with the metric induced from d_M. In §3.5 we encounter a continuous function $f : A \to N$ and we wish to extend it to a continuous function $g : M \to N$. When f is uniformly continuous, we can make this extension if we assume that N is complete and that points in M can be "approximated" by points in A in the following sense: A subset $A \subset M$ is called *dense* if $\bar{A} = M$.

Theorem 14.   Let (M, d_M) and (N, d_N) be metric spaces. Let $A \subset M$ be a dense subset of M, and let $f : A \to N$ be uniformly continuous. Assume that N is complete. Then there exists a unique continuous map $g : M \to N$ such that

$$g(x) = f(x) \qquad \text{for all } x \in A$$

Proof.   We begin by defining g. Let $x \in M$. If $x \in A$ we set $g(x) = f(x)$. If $x \notin A$ then, because A is dense in M, x must be a limit point of A. Choose a sequence $x_1, x_2, \ldots$ in A with $x_i \to x$ in M as $i \to \infty$. By proposition 13 the sequence $f(x_i)$ in N is Cauchy. Because N is assumed to be complete, we know that this sequence has a limit; define this limit to be $g(x)$. Notice that this definition is independent of the choice of sequence $x_1, x_2, \ldots$ because, if $x'_1, x'_2, \ldots$ is another such sequence, then $f(x'_i) \to g(x)$. In this fashion it is also easy to see that, given $\varepsilon > 0$, there exists a $\delta > 0$ such that, if $d_M(x, y) < \delta$ for $x \in M$ and $y \in A$, then $d_N[g(x), g(y)] < \varepsilon$. To prove continuity of g, take $x, y \in M$ with $d_M(x, y) < \delta/2$. Because A is dense in M, we can find $z \in A$ with $d_M(x, z) < \delta/2$. Then $d_M(y, z) \leq d_M(y, x) + d_M(x, z) < \delta$ and so

$$d_N[g(x), g(y)] \leq d_N[g(x), g(z)] + d_N[g(z), g(y)] < 2\varepsilon$$

whenever $d_M(x, y) < \delta/2$.

The uniqueness of g is a direct consequence of its continuity. $\square$

Appendix B
On $\mathscr{L}^p$ Matters

You recall that in §3.8 we proved the following improved version of the law of large numbers.

Theorem. Let $f_1, f_2, \ldots$ be bounded, independent, identically distributed random variables on the probability space (X, μ). Let E be the common expectation value of the f_i's, and let $S_n(x) = f_1(x) + \cdots + f_n(x) - nE$. Then, for any $\alpha > 0$,

$$(1) \qquad \frac{S_n(x)}{n^{(1/2)+\alpha}} \to 0$$

as $n \to \infty$ with probability one.

To prove this theorem we assumed that the f_i's were bounded so that we wouldn't have to worry about the integrability of the functions S_n^{2k}. Actually, if $\alpha > 1/k$ it is easy to see that equation 1 holds as long as

$$(2) \qquad \int |f_i|^{2k} < \infty$$

Indeed, in order to get the estimate in inequality 4 of §3.8, all you need to know is that

$$(3) \qquad \int |f_{i_1}^{l_1} f_{i_2}^{l_2} \times \cdots \times f_{i_j}^{l_j}| < \infty \qquad \text{when } l_1 + \cdots + l_j = 2k$$

It turns out that inequality 3 follows from inequality 2 once we have some basic facts about $\mathscr{L}^p$-spaces, which you have already proven in the exercises of §3.1 (see corollary 5 of this appendix). Here we first review those basic facts (don't peek until you've looked at exercises 7, 8, and 9 of §3.1 and exercise 10 of §3.2) and then we develop some $\mathscr{L}^p$ analogues of some of the $\mathscr{L}^2$-theory in Chapter 3.

Basic Theory

Let (X, μ) be a measure space. Recall that, for $p \geq 1$, $\mathscr{L}^p(X, \mu)$ (or just $\mathscr{L}^p$ if X, μ is understood) is the set of complex-valued measurable functions $f: X \to \mathbf{C}$ such that

$$(4) \qquad \qquad \int_X |f|^p \, d\mu < \infty$$

The value

$$(5) \qquad \qquad \| f \|_p = \left(\int_X |f|^p \, d\mu \right)^{1/p} \qquad f \in \mathscr{L}^p$$

is called the $\mathscr{L}^p$-norm of f.

Theorem 1. $\mathscr{L}^p(X, \mu)$ is a vector space and $\| \cdot \|_p$ is a norm on $\mathscr{L}^p(X, \mu)$.
   To prove theorem 1 we will need the following lemma of calculus.

Lemma 2.   Let $\phi(t)$ be a convex function on the interval (a, b); that is, $\phi''(t) \geq 0$ for all $t \in (a, b)$. Let $x, y \in (a, b)$ and let α and β be nonnegative numbers with $\alpha + \beta = 1$. Then

$$\phi(\alpha x + \beta y) \leq \alpha \phi(x) + \beta \phi(y)$$

Proof.   Suppose this lemma is not true. Then, for some α, β, x, and y as above, we have

$$\phi(\alpha x + \beta y) > \alpha \phi(x) + \beta \phi(y)$$

or

$$\alpha[\phi(\alpha x + \beta y) - \phi(x)] > \beta[\phi(y) - \phi(\alpha x + \beta y)]$$

because $\alpha + \beta = 1$. Dividing by $\alpha \beta(y - x)$, we get

$$\frac{\phi(\alpha x + \beta y) - \phi(x)}{(\alpha x + \beta y) - x} > \frac{\phi(y) - \phi(\alpha x + \beta y)}{y - (\alpha x + \beta y)}$$

By the mean value theorem there exist ξ and η with $x < \xi < \alpha x + \beta y < \eta < y$ such that $\phi'(\xi) > \phi'(\eta)$. This contradicts $\phi''(t) > 0$. ∇

Proof of theorem 1. First, to check that $\mathscr{L}^p(X, \mu)$ is a vector space, we need to see that, if $f, g \in \mathscr{L}^p(X, \mu)$, then $\int |f + g|^p < \infty$. To do this, notice that for $p \geq 1$ the function $\phi(t) = t^p$ is convex; hence, with $\alpha = \beta = \frac{1}{2}$, we conclude from lemma 2 that

$$\left(\frac{1}{2}|f| + \frac{1}{2}|g| \right)^p \leq \frac{1}{2}|f|^p + \frac{1}{2}|g|^p$$

pointwise in X. Integrating this inequality gives

$$\int |f + g|^p \, d\mu \leq 2^p \int \left(\frac{1}{2}|f| + \frac{1}{2}|g| \right)^p d\mu \leq 2^{p-1} \left(\int |f|^p \, d\mu + \int |g|^p \, d\mu \right)$$

Hence, $f + g \in \mathscr{L}^p(X, \mu)$.

To show that $\| \cdot \|_p$ is a norm on $\mathscr{L}^p(X, \mu)$, we need to prove the triangle inequality; that is, $\| f + g \|_p \leq \| f \|_p + \| g \|_p$. (The rest of the norm properties are obvious.) To prove this inequality, we need the following.

Lemma 3. Let f and g be nonnegative measurable functions on X, and let p and q be numbers greater than 1 with $(1/p) + (1/q) = 1$. Then

(6)
$$\int fg \, d\mu \leq \left(\int f^p \, d\mu \right)^{1/p} \left(\int g^q \, d\mu \right)^{1/q}$$

Proof. Let a and b be positive numbers. Define the numbers x and y by $a = e^{x/p}$ and $b = e^{y/q}$. Then, because e^t is convex and $(1/p) + (1/q) = 1$, we have from lemma 2 that

$$ab = e^{(x/p)+(y/q)} \leq \frac{1}{p}e^x + \frac{1}{q}e^y = \frac{1}{p}a^p + \frac{1}{q}b^q$$

Now let

$$a = \frac{f(x)}{\left(\int f^p \, d\mu \right)^{1/p}} \quad \text{and} \quad b = \frac{g(x)}{\left(\int g^q \, d\mu \right)^{1/q}}$$

and integrate to get inequality 6. $\triangledown$

To prove the triangle inequality, consider

$$\int |f + g|^p \, d\mu \leq \int (|f| + |g|)^p \, d\mu$$

$$= \int |f|(|f| + |g|)^{p-1} \, d\mu + \int |g|(|f| + |g|)^{p-1} \, d\mu$$

Apply lemma 3 with $q = p/(p-1)$ to get

$$\int (|f| + |g|)^p \, d\mu \leq \left(\int |f|^p \, d\mu \right)^{1/p} \left(\int (|f| + |g|)^p \, d\mu \right)^{(p-1)/p}$$

$$+ \left(\int |g|^p \, d\mu \right)^{1/p} \left(\int (|f| + |g|)^p \, d\mu \right)^{(p-1)/p}$$

so

(7) $$\|f + g\|_p \leq \left(\int (|f| + |g|)^p \, d\mu \right)^{1/p} \leq \|f\|_p + \|g\|_p \qquad \square$$

Remark.   The triangle inequality for $\| \cdot \|_p$, inequality 7, is called *Minkowski's inequality.*

We can now give an argument to show that inequality 2 implies inequality 3. First, we have the following proposition.

Proposition 4.   (Hölder's inequality) Let p and q be numbers greater than 1 with $(1/p) + (1/q) = 1$. Let $f \in \mathscr{L}^p(X, \mu)$ and $g \in \mathscr{L}^q(X, \mu)$. Then $fg \in \mathscr{L}^1(X, \mu)$ and

$$\left| \int fg \, d\mu \right| \leq \|f\|_p \|g\|_q$$

Proof.   By lemma 3 we have

$$\int |fg| \, d\mu = \int |f||g| \, d\mu \leq \|f\|_p \|g\|_p$$

so $fg \in \mathscr{L}^1(X, \mu)$ and

$$\left| \int fg \, d\mu \right| \leq \int |fg| \, d\mu \leq \|f\|_p \|g\|_q \qquad \square$$

Corollary 5.   Let p_1 and p_2 be greater than 1, and let $f_1 \in \mathscr{L}^{p_1}(X, \mu)$ and $f_2 \in \mathscr{L}^{p_2}(X, \mu)$. Then $f_1 f_2 \in \mathscr{L}^{p_1 p_2/(p_1 + p_2)}(X, \mu)$ and

$$\|f_1 f_2\|_{p_1 p_2/(p_1 + p_2)} \leq \|f_1\|_{p_1} \|f_2\|_{p_2}$$

Proof.   Consider

$$\int |f_1 f_2|^{p_1 p_2/(p_1 + p_2)} \, d\mu = \int |f_1|^{p_1 p_2/(p_1 + p_2)} |f_2|^{p_1 p_2/(p_1 + p_2)} \, d\mu$$

Notice that $f_1 \in \mathscr{L}^{p_1}(X, \mu)$ implies that $f_1^{p_1 p_2/(p_1 + p_2)} \in \mathscr{L}^{(p_1 + p_2)/p_2}$, and, similarly, $f_2^{p_1 p_2/(p_1 + p_2)} \in \mathscr{L}^{(p_1 + p_2)/p_1}$. Now $[p_2/(p_1 + p_2)] + [p_1/(p_1 + p_2)] = 1$, so we can apply proposition 4 to conclude that

$$\int (|f_1||f_2|)^{p_1 p_2/(p_1 + p_2)} \, d\mu \leq \left(\int |f_1|^{p_1} \, d\mu \right)^{p_2/(p_1 + p_2)} \left(\int |f_2|^{p_2} \, d\mu \right)^{p_1/(p_1 + p_2)}$$

so

$$\| f_1 f_2 \|_{p_1 p_2/(p_1 + p_2)} \leq \| f_1 \|_{p_1} \| f_2 \|_{p_2} \qquad \square$$

Now take k functions in $\mathscr{L}^k$, $f_1, f_2, \ldots, f_k$. By corollary 5, $f_1 f_2 \in \mathscr{L}^{k/2}$. Applying corollary 5 again gives $f_1 f_2 f_3 = (f_1 f_2) f_3 \in \mathscr{L}^{k/3}$. Continuing in this fashion, we get $f_1 f_2 \times \cdots \times f_k \in \mathscr{L}^1$; hence inequality 2 implies inequality 3.

Notice also that corollary 5 gives the following.

Corollary 6. Suppose $\mu(X) < \infty$. Then $\mathscr{L}^p(X, \mu) \subset \mathscr{L}^r(X, \mu)$ for $1 \leq r \leq p$ and

(8) $$\| f \|_r \leq c \| f \|_p \qquad \text{for } f \in \mathscr{L}^p(X, \mu)$$

where $c = [\mu(X)]^{(p-r)/pr}$.

Proof. Because $\mu(X) < \infty$ the constant function 1 is in $\mathscr{L}^s(X, \mu)$ for all $s \geq 1$. Let $s = pr/(p - r)$. Then, by corollary 5, $f \in \mathscr{L}^{ps/(p+s)} = \mathscr{L}^r$ and

$$\| f \|_r \leq \| f \|_p \| 1 \|_s \qquad \square$$

To continue now with the basic properties of $\mathscr{L}^p$-spaces, recall that a normed vector space is called a *Banach space* if it is complete in the metric topology.

Theorem 7. $\mathscr{L}^p(X, \mu)$ is a Banach space.

Proof. We will prove this theorem in the case that X is σ-finite; that is, $X = \bigcup_{i=1}^{\infty} X_i$ with $X_1 \leq X_2 \leq \cdots$ and $\mu(X_i) < \infty$ for all i. We leave the more general case to the reader (see exercises 8 and 9 in §3.2).

Let $\{ f_n \}_{n=1}^{\infty}$ be a Cauchy sequence in $\mathscr{L}^p(X, \mu)$; that is, given $\varepsilon > 0$ there is an N such that $\| f_n - f_m \|_p < \varepsilon$ whenever $n, m > N$. We wish to show that there is a function $f \in \mathscr{L}^p(X, \mu)$ such that $f_n \to f$ in $\mathscr{L}^p$; that is, given $\varepsilon > 0$ there is an N such that $\| f_n - f \|_p < \varepsilon$ when $n > N$.

We proceed as in the proof of theorem 13 of §3.2. Because $\mu(X_1) < \infty$ we get from corollary 5 that $\mathscr{L}^p(X_1, \mu) \subset \mathscr{L}^1(X_1, \mu)$, and the f_n's form a Cauchy sequence in $\mathscr{L}^1(X_1, \mu)$. Thus, from theorem 9 of §3.1, we can extract a subsequence $\{ f_{1,n} \}_{n=1}^{\infty}$ that converges a.e. on X_1. Similarly, because $\mu(X_2) < \infty$

we can extract a subsequence $\{f_{2,n}\}_{n=1}^{\infty}$ from the sequence $\{f_{1,n}\}_{n=1}^{\infty}$ that converges a.e. on X_2. Continuing inductively we extract a subsequence $\{f_{i,n}\}_{n=1}^{\infty}$ from $\{f_{i-1,n}\}_{n=1}^{\infty}$ that converges a.e. on X_i. By the Cantor diagonal process, the subsequence $f_{1,1}, f_{2,2}, \ldots$ converges a.e. on X to a measurable function f on X. Let $g_1 = f_{1,1}$, $g_2 = f_{2,2}$, and so on. The sequence $\{g_n\}_{n=1}^{\infty}$ is Cauchy in $\mathscr{L}^p(X, \mu)$; that is, for $\varepsilon > 0$ there is an N such that $\|g_m - g_n\|_p^p < \varepsilon$ when $m, n > N$.

By Fatou's lemma, with $n > N$ fixed and $m \to \infty$,

$$\int \liminf |g_m - g_n|^p \, d\mu \le \liminf \int |g_m - g_n|^p \, d\mu \le \varepsilon$$

The term on the left is $\int |f - g_n|^p \, d\mu$, so this inequality shows that $f \in \mathscr{L}^p(X, \mu)$ and that $g_n \to f$ in $\mathscr{L}^p(X, \mu)$. Because $\{g_n\}_{n=1}^{\infty}$ is a subsequence of $\{f_n\}_{n=1}^{\infty}$, it follows that $f_n \to f$ in $\mathscr{L}^p$ as well. $\square$

Representation Theorems

Recall that, for the Hilbert space $\mathscr{L}^2(X, \mu)$, we have Schwarz's inequality

$$\left| \int fg \, d\mu \right| \le \|f\|_2 \|g\|_2 \qquad \text{for } f, g \in \mathscr{L}^2(X, \mu)$$

The generalization of this inequality to $\mathscr{L}^p$-spaces is Hölder's inequality

$$\left| \int fg \, d\mu \right| \le \|f\|_p \|g\|_q, \qquad f \in \mathscr{L}^p(X, \mu), \qquad g \in \mathscr{L}^q(X, \mu), \qquad \frac{1}{p} + \frac{1}{q} = 1$$

This inequality can be interpreted in the following way: For $g \in \mathscr{L}^q(X, \mu)$ consider the linear map

(9) $l_g : \mathscr{L}^p(X, \mu) \to \mathbf{C}$ where $l_g(f) = \int fg \, d\mu$

It is an immediate consequence of Hölder's inequality that this map is continuous. Indeed, to show that a map $F : \mathscr{L}^p(X, \mu) \to \mathbf{C}$ is continuous at f_0, we need to show that, given $\varepsilon > 0$, there is a $\delta > 0$ such that $\|f - f_0\|_p < \delta$ implies that $|F(f) - F(f_0)| < \varepsilon$. In this case

$$|l_g(f) - l_g(f_0)| = |l_g(f - f_0)| \le \|f - f_0\|_p \|g\|_q$$

so, if we take $\delta = \varepsilon / \|g\|_q$, we are done.

The purpose of this section is to convince you that, in fact, every continuous linear map $l : \mathscr{L}^p(X, \mu) \to \mathbf{C}$ is of the form l_g for some $g \in \mathscr{L}^q(X, \mu)$ where $(1/p) + (1/q) = 1$. We will not prove this fact in general; we will just prove some special cases of it. If you want to see the general proofs, we recommend the treatment in Reed and Simon (*Methods of Modern Mathematical Physics: Functional Analysis*, vol. 1. New York: Academic Press, 1980).

We first introduce some nomenclature. Let $(V, \|\cdot\|)$ be a normed, complex vector space. A continuous linear map $l : V \to \mathbf{C}$ is called a *continuous linear functional*. The *dual space* V^* is the space of all continuous linear functionals $l : V \to \mathbf{C}$. Notice that V^* is a vector space.

Proposition 8. Let $l : V \to \mathbf{C}$ be a linear functional. l is continuous if and only if there is a constant $c > 0$ such that

$$(10) \qquad\qquad l(v) \le c\|v\| \qquad \text{for all } v \in V$$

Proof. The argument that inequality 10 implies continuity is the same as the argument that l_g is continuous. To show that continuity of l implies inequality 10, notice that, because l is continuous at zero, if we choose $\varepsilon > 0$ there is a $\delta > 0$ such that $\|v\| < \delta$ implies $|l(v)| < \varepsilon$. Let $c = 2\varepsilon/\delta$. Then for $v \in V$ notice that

$$\left\| \frac{\delta v}{2\|v\|} \right\| < \delta \qquad \text{so} \qquad \left| l\left(\frac{\delta v}{2\|v\|} \right) \right| = \frac{\delta}{2\|v\|} |l(v)| < \varepsilon$$

Hence $|l(v)| < 2\varepsilon/\delta \|v\| = c\|v\|$. $\square$

Now let H be a Hilbert space with inner product $\langle \cdot, \cdot \rangle$. If $v \in H$ we define $l_v \in H^*$ by $l_v(w) = \langle w, v \rangle$, $w \in H$. (l_v is continuous by Schwarz's inequality.) This definition gives a map

$$L : H \to H^* \qquad \text{by } L(v) = l_v$$

It is easy to see that L is one to one. Indeed, if $L(v) = 0$ then $l_v(w) = \langle v, w \rangle = 0$ for all w; hence $v = 0$. The surprising fact is that L is onto.

Theorem 9. The map $L : H \to H^*$ defined by $L(v) = l_v$ is bijective.

Remark. We call this theorem a representation theorem because it "represents" the abstract space H^* in terms of the known space H.

To prove theorem 9 we will need a geometric result on Hilbert spaces, which is interesting in its own right. Let $V \subset H$ be a vector subspace of H. If V is also a closed subset of H (in the norm topology), then it is automatically a Hilbert space itself; in this case V is called a *Hilbert subspace* of H. If $w \in H$ we write $w \perp V$ if $\langle w, v \rangle = 0$ for all $v \in V$. Let $V^{\perp}$ be the set of all $w \in H$ such that $w \perp V$. Notice that $V^{\perp}$ is a vector subspace of H.

Proposition 10. Let $V \subset H$ be a Hilbert subspace of the Hilbert space H. Assume that $V \ne H$. Then there exists $w \in V^{\perp}$ such that $w \ne 0$.

Proof.   Because $V \neq H$ there is an $x \in H$ such that $x \notin V$. Let

$$a = \inf_{v \in V} \|x - v\|$$

We claim that there is a $y \in V$ such that $\|x - y\| = a$. To see this fact choose $v_n \in V$ such that $\|x - v_n\| \to a$ as $n \to \infty$. Then

$$\|v_n - v_m\|^2 = 2\|v_n - x\|^2 + 2\|v_m - x\|^2 - 4\|\tfrac{1}{2}(v_n + v_m) - x\|^2$$

$$\leq 2\|v_n - x\|^2 + 2\|v_m - x\|^2 - 4a^2$$

Hence the v_n's form a Cauchy sequence in H. Because V is closed in H, the sequence of v_n's converges to $y \in V$ and

$$\|x - y\| = \lim_{n \to \infty} \|x - v_n\| = a$$

Now let $w = x - y$, $w \neq 0$, because $y \in V$ and $x \notin V$. We claim that $w \in V^\perp$. If this claim is not true, then there is a $v_0 \in V$ such that $\langle w, v_0 \rangle \neq 0$. Let

$$y' = y + \frac{\langle w, v_0 \rangle}{\|v_0\|^2} v_0$$

Then $y' \in V$ and

$$\|x - y'\|^2 = \left\| x - y - \frac{\langle w, v_0 \rangle}{\|v_0\|^2} v_0 \right\|^2$$

$$= \|x - y\|^2 + \frac{|\langle w, v_0 \rangle|^2}{\|v_0\|^2} - \frac{\langle w, v_0 \rangle}{\|v_0\|^2} \langle x - y, v_0 \rangle - \frac{\overline{\langle w, v_0 \rangle}}{\|v_0\|^2} \langle v_0, x - y \rangle$$

$$= \|x - y\|^2 - \frac{|\langle w, v_0 \rangle|^2}{\|v_0\|^2}$$

$$< \|x - y\|^2$$

This inequality contradicts the fact that $\|x - y\| = a$.   □

We now prove theorem 9. Let $l \in H^*$. We want to find $v \in H$ such that $l = l_v$. If $l \equiv 0$ we can take $v = 0$, so we will assume $l \not\equiv 0$ from now on. Let $K \subset H$ be the subspace of H defined by

$$K = \{w \in H; \, l(w) = 0\}$$

Note that $K \neq H$ because $l \not\equiv 0$. If $\{v_n\}$ is a Cauchy sequence in K, then $v_n \to v$ for some $v \in H$. By the continuity of l, we see that $l(v_n) \to l(v)$ so $l(v) = 0$. Hence $v \in K$. Thus we have seen that K is closed. From proposition 9 there exists $w \neq 0$ in $K^\perp$. Let

$$v = \frac{\overline{l(w)}}{\|w\|^2} w$$

Then

$$l(v) = \frac{|l(w)|^2}{\|w\|^2} = \|v\|^2$$

Let x be any element of H. Notice that

$$l\left(x - \frac{l(x)}{\|v\|^2}v\right) = l(x) - l(x) = 0$$

so

$$x - \frac{l(x)}{\|v\|^2}v \in K$$

But $v \in K^\perp$, so

$$0 = \left\langle x - \frac{l(x)}{\|v\|^2}v, v \right\rangle = \langle x, v \rangle - l(x)$$

That is, $l(x) = \langle x, v \rangle$ for all $x \in H$.   □

Now let's return to $\mathscr{L}^p$-spaces. In the beginning of this section, we showed that, if $g \in \mathscr{L}^q$ with $(1/p) + (1/q) = 1$, then g defines a continuous linear functional $l_g \in \mathscr{L}^{p*}$ given by $l_g(f) = \int fg\,d\mu$ for $f \in \mathscr{L}^p$. This definition gives a map

$$L : \mathscr{L}^q \to \mathscr{L}^{p*} \qquad \text{by } L(g) = l_g$$

This map is one to one because if $l_g \equiv 0$ then, in particular, $l_g(1_A) = \int_A g\,d\mu = 0$ for all measurable sets A with $\mu(A) < \infty$. This implies $g = 0$ a.e.

In fact, when X is σ-finite, it is also true that L is onto; that is, L is an isomorphism. We won't prove this in full generality; instead we'll prove the following special case.

Theorem 11. Assume that $\mu(X) < \infty$ and let $1 < p \leq 2$ and $q = p/(p-1)$. Then $L : \mathscr{L}^q \to \mathscr{L}^{p*}$ is an isomorphism.

Proof. Let $l \in \mathscr{L}^{p*}$. We want to find $g \in \mathscr{L}^q$ such that $l = l_g$. Recall that, because $\mu(X) < \infty$, corollary 6 implies that $\mathscr{L}^2(X, \mu) \subset \mathscr{L}^p(X, \mu)$ and that there is a constant $a > 0$ such that

$$\|f\|_p \leq a\|f\|_2 \qquad \text{for } f \in \mathscr{L}^2$$

Now, by proposition 8, there is a constant $c > 0$ such that

$$|l(f)| \leq c\|f\|_p \qquad \text{for } f \in \mathscr{L}^p$$

Thus

$$|l(f)| \le c\|f\|_p \le ca\|f\|_2 \qquad \text{for } f \in \mathscr{L}^2$$

Applying proposition 8 again we see that, if we restrict l to $\mathscr{L}^2$, we get something in $\mathscr{L}^{2*}$. By theorem 9 there is a $g \in \mathscr{L}^2$ such that

$$l(f) = \int f\bar{g}\,d\mu \qquad \text{for } f \in \mathscr{L}^2$$

We claim that g is actually in $\mathscr{L}^q$. To see this, fix $K > 0$ and set

$$h(x) = \begin{cases} |g(x)|^{q-1} & \text{if } |g(x)| \le K \\ 0 & \text{if } |g(x)| > K \end{cases}$$

Then $h(x)$ is bounded, so $h \in \mathscr{L}^2(X, \mu)$ and

$$|l(h)| = \int_{|g(x)| \le K} |g|^q\,d\mu \le c\|h\|_p$$

$$= c\left(\int_{|g(x)| \le K} |g(x)|^{(q-1)p}\,d\mu\right)^{1/p}$$

$$= c\left(\int_{|g(x)| \le K} |g(x)|^q\,d\mu\right)^{1-(1/q)}$$

So

$$\left(\int_{|g(x)| \le K} |g(x)|^q\,d\mu\right)^{1/q} \le c$$

This bound holds for all K, so, by the monotone convergence theorem, we conclude that $g(x) \in \mathscr{L}^q$.

Finally, because $\mathscr{L}^2(X, \mu)$ is dense in $\mathscr{L}^p(X, \mu)$ (e.g., the simple functions are dense in $\mathscr{L}^p$), it is clear by continuity that

$$l(f) = \int f\bar{g}\,d\mu \qquad \text{for all } f \in \mathscr{L}^p(X, \mu) \qquad\qquad \square$$

Remark.   It is an easy exercise to extend this result to the case when X is σ-finite.

Convolution

In the exercises of §3.5, we defined the *convolution* of f and g when f and g are in $\mathscr{L}^1(\mathbf{R})$ by

(11)
$$f * g(x) = \int_{\mathbf{R}} f(x - y)g(y)\,dy$$

In those exercises you proved the following proposition.

Proposition 12. $f * g \in \mathscr{L}^1(\mathbf{R})$ and

$$\|f * g\|_1 \leq \|f\|_1 \|g\|_1$$

Proof. Consider the iterated integral

$$\int_{\mathbf{R}} \int_{\mathbf{R}} |f(x - y)g(y)|\,dx\,dy = \|f\|_1 \|g\|_1$$

By Fubini's theorem (theorem 15 of §2.5), the integral

$$\int_{\mathbf{R}} f(x - y)g(y)\,dy$$

makes sense for almost all x and is equal a.e. to an $\mathscr{L}^1$ function. Thus $f * g \in \mathscr{L}^1(\mathbf{R})$ and a second application of Fubini's theorem gives

$$\|f * g\|_1 = \int_{\mathbf{R}} \left| \int_{\mathbf{R}} f(x - y)g(y)\,dy \right| dx$$

$$\leq \int_{\mathbf{R}} \int_{\mathbf{R}} |f(x - y)g(y)|\,dy\,dx$$

$$= \int_{\mathbf{R}} \int_{\mathbf{R}} |f(x - y)g(y)|\,dx\,dy$$

$$= \|f\|_1 \|g\|_1 \qquad \square$$

Corollary 13. Convolution is a continuous map from $\mathscr{L}^1 \times \mathscr{L}^1 \to \mathscr{L}^1$; that is, if $f_n \to f$ in $\mathscr{L}^1$ and $g_n \to g$ in $\mathscr{L}^1$, then $f_n * g_n \to f * g$ in $\mathscr{L}^1$.

Proof. Choose n large enough so that $\|f - f_n\|_1 < 1$. Then $\|f_n\|_1 \leq (1 + \|f\|_1)$. Now

$$\|f * g - f_n * g_n\|_1 = \|(f - f_n) * g + f_n * (g - g_n)\|_1$$

$$\leq \|f - f_n\|_1 \|g\|_1 + \|f_n\|_1 \|g - g_n\|_1$$

$$\leq \|f - f_n\|_1 \|g\|_1 + (1 + \|f\|_1) \|g - g_n\| \to 0 \quad \text{as} \quad n \to \infty$$

$$\square$$

Proposition 14. Let $f, g, h \in \mathcal{L}^1(\mathbf{R})$. Then

1. $f * g = g * f$
2. $(f * g) * h = f * (g * h)$

Proof.

1. $f * g(x) = \displaystyle\int_{\mathbf{R}} f(x - y)g(y)\,dy$

   $= \displaystyle\int_{\mathbf{R}} f(s)g(x - s)\,ds, \qquad s = x - y$

   $= g * f(x)$

2. $(f * g) * h(x) = \displaystyle\int_{\mathbf{R}}\int_{\mathbf{R}} f(x - z - y)g(y)\,dy\,h(z)\,dz$

   $= \displaystyle\int_{\mathbf{R}}\int_{\mathbf{R}} f(x - s)g(s - z)\,ds\,h(z)\,dz, \qquad s = z + y$

   $= \displaystyle\int_{\mathbf{R}} f(x - s)\int_{\mathbf{R}} g(s - z)h(z)\,dz\,ds$

   $= f * (g * h)(x)$ □

One of the main strengths of the convolution is that it tells us what corresponds to a product in Fourier transform land.

Proposition 15. $(f * g)\,\hat{} = \hat{f}\hat{g}$ for $f, g \in \mathcal{L}^1(\mathbf{R})$.

Proof. $(f * g)\,\hat{}(\xi) = \displaystyle\int_{\mathbf{R}} e^{-ix\xi}\int_{\mathbf{R}} f(x - y)g(y)\,dy\,dx$

$= \displaystyle\int_{\mathbf{R}}\int_{\mathbf{R}} e^{-ix\xi}f(x - y)g(y)\,dx\,dy$

$= \displaystyle\int_{\mathbf{R}}\int_{\mathbf{R}} e^{-i(s+y)\xi}f(s)g(y)\,ds\,dy$

$= \hat{f}(\xi)\displaystyle\int_{\mathbf{R}} e^{-iy\xi}g(y)\,dy$

$= \hat{f}(\xi)\hat{g}(\xi)$ □

The convolution can be extended to $\mathcal{L}^p$-spaces in various combinations. For instance, if $f \in \mathcal{L}^p$ and $g \in \mathcal{L}^q$ for $(1/p) + (1/q) = 1$, then $f(x - y)g(y) \in \mathcal{L}^1$

for each fixed x; so equation 11 makes sense. In fact by Hölder's inequality we have

(12) $|f * g(x)| \leq \| f \|_p \| g \|_q$ for all x

Hence, $f * g$ is a bounded function when $f \in \mathscr{L}^p$ and $g \in \mathscr{L}^q$, $(1/p) + (1/q) = 1$.

Proposition 16.   Let $f \in \mathscr{L}^p$, $g \in \mathscr{L}^q$, and $(1/p) + (1/q) = 1$. Then $f * g$ is a bounded, uniformly continuous function on **R**.

Proof.   Let $\varepsilon > 0$ be given. We need to find $\delta > 0$ such that if $|x - y| < \delta$ then $|f * g(x) - f * g(y)| < \varepsilon$. Notice that

$$|f * g(x) - f * g(y)| = \left| \int [f(x - z) - f(y - z)]g(z)\, dz \right|$$

$$\leq \| f_x - f_y \|_p \| g \|_q$$

where $f_x(z) = f(x - z)$. Now let ϕ be a compactly supported continuous function with

$$\| f - \phi \|_p < \varepsilon$$

(You can prove that such ϕ's exist.) It is easy to see that there is a $\delta > 0$ such that if $|x - y| < \delta$ then

$$\| \phi_x - \phi_y \|_p < \varepsilon$$

Then $\| f_x - f_y \|_p \leq \| f_x - \phi_x \|_p + \| \phi_x - \phi_y \|_p + \| f_y - \phi_y \|_p < 3\varepsilon$ □
    Another combination for which the convolution can be defined is $f \in \mathscr{L}^p$ and $g \in \mathscr{L}^1$.

Proposition 17.   Let $f \in \mathscr{L}^p$, $1 \leq p$, and $g \in \mathscr{L}^1$. Then $f * g$ is well-defined and is in $\mathscr{L}^p$. Moreover

$$\| f * g \|_p \leq \| f \|_p \| g \|_1$$

Proof.   Let $q = p/(p - 1)$ and notice that

$$\int |f(x - y)g(y)|\, dy = \int |g(y)|^{1/p} |f(x - y)| |g(y)|^{1/q}\, dy$$

Now, by proposition 12, $|g(y)| |f(x - y)|^p$ is integrable for almost all x, because $|g|$ and $|f|^p$ are both in $\mathscr{L}^1$. Thus, $|g(y)|^{1/p} |f(x - y)|$ is in $\mathscr{L}^p$ for almost all x. Hence, for almost all x we have by Hölder's inequality

$$\int |f(x - y)g(y)|\, dy \leq \left(\int |g(y)| |f(x - y)|^p\, dy \right)^{1/p} \| g \|_1^{1/q}$$

Thus we have that $f * g$ is defined for almost all x; furthermore, we get

$$|f * g(x)|^p \le \left(\int |g(y)| \, |f(x - y)|^p \, dy \right) \|g\|_1^{p/q}$$

By Fubini's theorem

$$\|f * g\|_p^p \le \left(\int\!\!\int |g(y)| \, |f(x - y)|^p \, dx \, dy \right) \|g\|_1^{p/q}$$

$$\le \|f\|_p^p \|g\|_1^{1+p/q}$$

so $$\|f * g\|_p \le \|f\|_p \|g\|_1 \qquad\qquad \square$$

Another important use of the convolution is that it allows us to give explicit smooth approximations to $\mathscr{L}^p$ functions. We describe this in the following.

Proposition 18.   Let $f \in \mathscr{L}^1$ and let ϕ be a Schwartz function. Then $\phi * f$ is C^∞.

Proof. $\dfrac{d^k}{dx^k} (\phi * f)(x) = \dfrac{d^k}{dx^k} \int \phi(x - y) f(y) \, dy$

$$= \int \frac{d^k}{dx^k} \phi(x - y) f(y) \, dy$$

where the differentiation under the integral can be justified by the dominated convergence theorem. $\square$

Now let $\phi_0(x)$ be a smooth function with support in $(-1, 1)$ and such that

$$\int_{\mathbf{R}} \phi_0(x) \, dx = 1$$

Let $\phi_k(x) = k\phi_0(kx)$, $k = 1, 2, \ldots$. Then $\phi_k(x)$ is smooth, supported in $(-1/k, 1/k)$, and

$$\int_{\mathbf{R}} \phi_k(x) \, dx = 1$$

Theorem 19.   Let $f \in \mathscr{L}^p$. Then $\phi_k * f$ converges to f in $\mathscr{L}^p$, as $k \to \infty$.

Proof.   Let g be a smooth, compactly supported function with $\|f - g\|_p < \varepsilon$. Then

$$\|\phi_k * f - f\|_p \le \|\phi_k * f - \phi_k * g\|_p + \|\phi_k * g - g\|_p + \|g - f\|_p$$

By proposition 17, $\|\phi_k * (f - g)\|_p \leq \|f - g\|_p$ because $\|\phi_k\|_1 = 1$. Thus

$$\|\phi_k * f - f\|_p \leq 2\varepsilon + \|\phi_k * g - g\|_p$$

Now notice that

$$|\phi_k * g(x) - g(x)| = \left| \int [g(x - y) - g(x)] \phi_k(y) \, dy \right|$$

because $\int \phi_k(y) \, dy = 1$. Because g is compactly supported, it is uniformly continuous; that is, given $\varepsilon > 0$ there is a $\delta > 0$ such that if $|y| < \delta$ then $|g(x - y) - g(y)| < \varepsilon$. Now ϕ_k is supported in $(-1/k, 1/k)$, so, if we take k large enough, we can make

$$|\phi_k * g(x) - g(x)| < \varepsilon \qquad \text{when } x \in \text{supp } g$$

Furthermore, if g is supported in (a, b), then $\phi_k * g$ is supported in $(a - 1/k, b + 1/k)$. Thus, for large enough k, we can conclude

$$\|\phi_k * g - g\|_p < \varepsilon \qquad \qquad \square$$

Fourier Transform in $\mathscr{L}^p(\mathbf{R})$

In §3.5 we defined the Fourier transform

$$\hat{f}(y) = \int f(x) e^{-ixy} \, dy \qquad \text{for } f \in \mathscr{L}^1(\mathbf{R})$$

We then used the density of the Schwartz space S in $\mathscr{L}^2$, along with the Plancherel formula, to extend the Fourier transform to an isomorphism of $\mathscr{L}^2$ onto $\mathscr{L}^2$. The same techniques can be used to extend the Fourier transform to $\mathscr{L}^p$ functions for $1 \leq p \leq 2$. We need the following two results.

Proposition 20. The Schwartz space S is dense in $\mathscr{L}^p(\mathbf{R})$, $p \geq 1$.

We leave the proof of this proposition to the reader as an exercise. It is essentially the same as the proof for $p = 2$ (theorem 15 of §3.5).

The other ingredient we need is an $\mathscr{L}^p$ replacement for the Plancherel formula.

Theorem 21. (Hausdorff–Young inequality) Let $1 < p \leq 2$ and $q = p/(p - 1)$. Then, for Schwartz functions f,

(13) $$\|\hat{f}\|_q \leq C \|f\|_p \qquad \text{for some constant } C$$

A good reference for the proof of this is the book by Reed and Simon (see the reference section, page 199).

This result tells us that the Fourier transform is uniformly continuous as a map from the dense set $S \subset \mathscr{L}^p(\mathbf{R})$ into the space $\mathscr{L}^q(\mathbf{R})$. By theorem 14 of Appendix A we can then extend it to a map of $\mathscr{L}^p(\mathbf{R})$ into $\mathscr{L}^q(\mathbf{R})$, and inequality 13 still holds for all $f \in \mathscr{L}^p(\mathbf{R})$.

References

P. Billingsley, *Probability and Measure*. New York: Wiley, 1979.

W. Feller, *An Introduction to Probability Theory and Its Applications, 3d ed.* New York: Wiley, 1967.

U. Grenander and G. Szegö, *Toeplitz Forms and Their Applications*. Berkeley, Calif.: University of California Press, 1958.

K. Hoffman, *Analysis in Euclidean Space*. Englewood Cliffs, N.J.: Prentice-Hall, 1975.

M. Kac, *Statistical Independence in Probability Analysis and Number Theory*. Mathematical Association of America (Carus Mathematical Monograph, no. 12) 1959.

S. Lang, *A Complete Course in Calculus*. Reading, Mass.: Addison-Wesley, 1968.

M. Reed and B. Simon, *Methods of Modern Mathematical Physics: Functional Analysis*, vol. 1, revised and enlarged edition. New York: Academic Press, 1980.

W. Rudin, *Principles of Mathematical Analysis, 3rd ed.* New York: McGraw-Hill, 1976.

Index